In situ and float

In situ + *float*

DR MARK BARSDELL

LEABROOK press

In situ and float

Leabrook Press
Adelaide, Australia
www.leabrookpress.com

Photographs © Mark Barsdell 2018
Cover design by Jeremy Majid
Editing by Wendy Barsdell
Proofreading by Susan Barsdell
Book design by Leabrook Press
Maps on pages 6 & 124 © OpenStreetMap contributors

Typeset in 11.5 / 17pt Crimson Text

ISBN: 978-0-9943700-3-7

NATIONAL LIBRARY OF AUSTRALIA
A catalogue record for this book is available from the National Library of Australia

LEABROOK press

To the explorers -
those who travel the world in the search for knowledge.

Contents

Foreword

My brother and I laid out four chairs, all facing the direction of the canvas screen. Familiar whirrs and clicks sounded behind us and suddenly the screen filled with colour. It was an unintelligible image. Some more whirrs and clicks, a curse word, and the image was righted – slide night could begin. From the perspective of a young imagination the slides were taken from the camera of a great explorer, a brave adventurer, an epic voyager, a pioneer of new lands. This was my father, but not as I knew him. The slides were from his expeditions and travels to remote places when he was studying and working in the field of geology; he had left geology for financial reasons before I was born.

Slide night would occur a few times per year and it was always highly anticipated. Dad would provide first person commentary, including descriptions of the weather conditions as well as geological features; the latter flying straight over our heads. He also answered our questions, such as "How did you go to the toilet?" "Could you have frozen to death going to the toilet?" and "Did everyone go to the toilet?" The slides reflected foreign lands, beyond any I had witnessed my father trek, but I reconciled the difference between my 'Important Office-worker Dad' and my 'Great Adventurer Dad' without question. In one corner of the lounge room (aka theatre) sat a glass cabinet filled with artefacts from his travels, as well as collections of shells and minerals. In the other corner lay a suit jacket and a sleek black briefcase with silver buckles.

I was aware my father kept diaries of his field trips and explorations. Rather than lengthy self-reflections or emotional confessions, I presumed them to contain detailed descriptions of the environments he traversed. The images that are interspersed throughout this book had been explained to me only in his commentary and in my imagination, but as my father's 70th birthday loomed in the year ahead I decided that it was time to learn more about what shaped the man I know today. With Mum assigned the role of undercover agent, my husband as project manager and my brother as a partner in crime, I accessed his diaries with the intention of forming them into this book.

As I read through the pages, I was privy to context that nobody had ever heard and his photos became even richer in my mind. He travelled to the far reaches of the earth – at a time when travel was a challenge in itself – and endured harsh conditions for prolonged periods. He put his safety at great risk on numerous occasions, making me astounded that I am here to publish his tale. Never once was I convinced that I could have done what he did, although perhaps that same thought formed *his* motivation – a personal drive that is apparent in his text.

I now have a family of my own and, looking back, while I idolised my father for his adventures and bravery, and I am enormously proud of his unique experiences, he was around for us and provided for us in ways an epic voyager couldn't have. I could be swept away in his photos, but appreciated him more than I realised *in situ*. Now in his retirement, his adventures have taken the form of holidays with my mother (albeit to places with at least one volcano as a minimum requirement) and new plans for his beloved garden. The man who survived uninhabitable conditions now chooses to enjoy his beverage on a plush sofa, rather than a crate at his local hipster café – despite the latter having far superior coffee I might add. This is not to say that I believe his pull to adventure has faded over time. Rather, that his values have evolved to accommodate creature comforts and time spent with his growing family. This is great comfort to me as it provides a degree of reassurance that there will not be a sequel. Nevertheless, his diaries capture a time when true adventure and exploration were possible and it is hoped that this book inspires renewed appreciation for the natural environment.

Wendy Barsdell

"Rocks must first be discovered, before they are analysed and finally understood. The geologist must therefore be an explorer, scientist and storyteller."

Anonymous

In situ
+
float

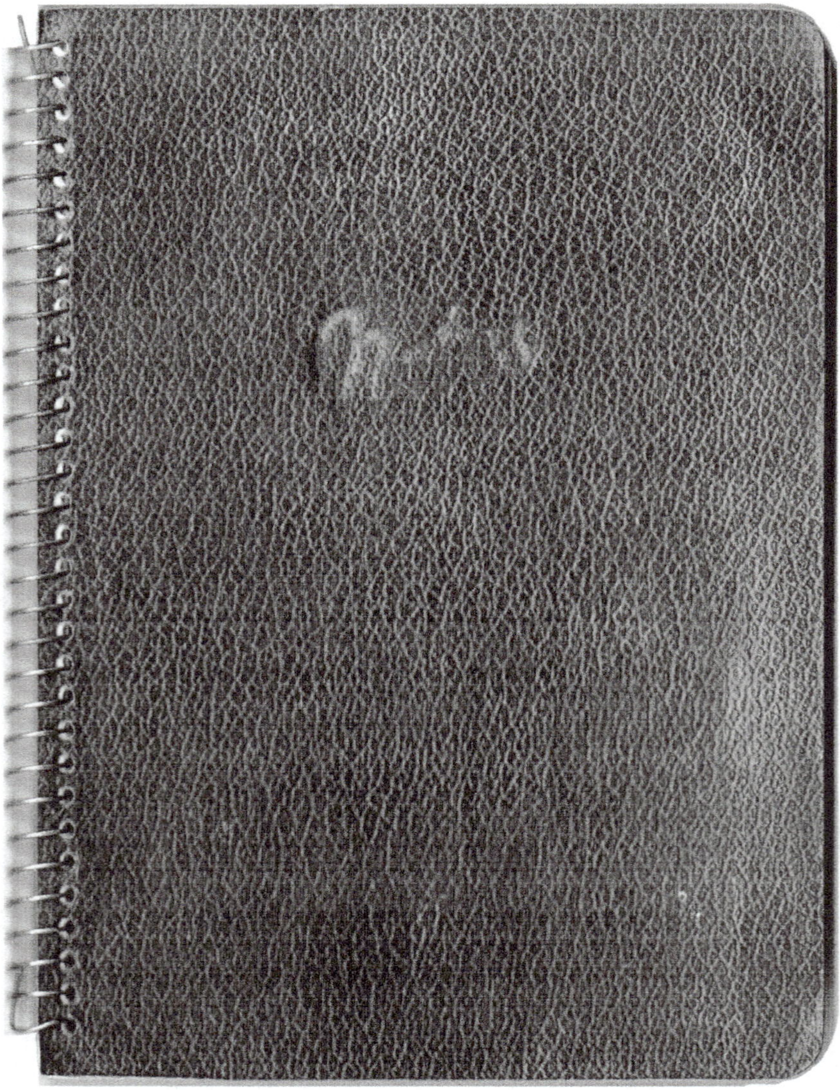

Introduction

For me it all started on August 21, 1948, in Melbourne. Only a few months later we left for Holland where my father took up a posting to work for the News and Information Bureau in The Hague, providing accurate information about Australia to foreigners. We remained in Holland for seven years before returning to Canberra, Australia.

Two years later my father was posted to Singapore where my brother and I attended the British army school. In 1960 we moved to Kuala Lumpur at a turbulent time - only three years after the Malayan states had gained independence from Great Britain (the states later became the federation of Malaysia 1963). I spent a year at school in Kuala Lumpur then moved to Sydney, Australia, for boarding school. Boarding school was a culture shock and a very unpleasant experience. Discipline was harsh and included frequent use of the cane and running around the oval every morning before breakfast, right through winter. I left after two years when my father was transferred back to Canberra and there I completed my last three years of high school.

My first year out of school was spent in the public service as a 'Junior Letter Writer', before completing a Bachelor of Science in Geology at the Australian National University. I worked a holiday job with Great Boulder Gold Mines near the small town of Kalgoorlie in Western Australia.

After completing my degree in 1969, I applied to the Bureau of Mineral Resources to work for five months. They offered me a position in New Guinea, based in Port Moresby. There, my accommodation was a nice little bungalow, not far from the water, with a communal dining room shared by other expats.

The geological work involved mapping part of the Owen Stanley Ranges, including sections of the Kokoda Track, the site of several bloody battles against the Japanese in WWII. My first field trip involved flying to a small town called Tetebedi in a single engine plane and then walking to Popondetta. I always remember my first night camping in the New Guinea bush, lying on a stretched canvas bed under a rain fly with open sides. There were a lot of sounds and I remember being quite nervous about some of them. My colleague and I were supported by about six local geological assistants who built the camps, carried the gear and prepared the evening meals.

I left New Guinea for New York via Sydney on a Qantas 707 to join my parents who were living on the corner of 1st Avenue. My father was running the Australian News and Information Bureau in New York. The Bureau mainly provided information for the U.S. government and corporates regarding Australia. His previous post involved running the London office. Shortly after arriving in New York I managed to get a job at Columbia University, working for Owen, the associate professor of structural geology who hailed from Scotland. The work mainly involved a petrographic analysis of a Tonalite intrusion on Livingstone Island in the South Shetlands. There was also an office at the Lamont Doherty Geological Observatory, which was situated on the Palisades Sill overlooking the Hudson River. The Lamont Doherty observatory is part of Columbia University and is famous for its ground-breaking work on identifying mid-ocean ridges and elucidating the theory of sea-floor spreading. I used to go over there on Fridays to attend the afternoon seminars.

It was through Owen that I would get my first opportunity to be part of a major field trip.

South America
South Orkney Islands
Antarctica

Antarctica

1971

Coronation Island
Powell Island
Laurie Island
Signy Island
SOUTH ORKNEY ISLANDS

18th-19th January 1971 - Age 22
New York/Miami/Panama

We left Kennedy Airport at 9pm on Braniff flight 979 (a Douglass DC8). There were four of us: Owen (the team leader), Phill (who had recently completed his PhD in structural geology), his wife Abby (a geology graduate) and myself (a recent graduate in geology). Over the previous few months, I had been doing petrographic (microscope interpretation) work for Owen on a suite of intrusive rocks (tonalites) from southern South America. The purpose of this trip was to untangle the structural history of the South Orkney Islands and to integrate our findings with the known structural patterns in the Antarctic Peninsula and southern South America, which were once joined. The South Orkney Islands at 60°S are much colder than the Orkney Islands at 60°N due to the warming effects of the Gulf Stream in the northern hemisphere.

We arrived at Washington DC around 10pm, remaining there for 25 minutes before flying on to Miami where we touched down at 1am in pouring rain. We stayed there for 1.5 hours then continued our journey south, arriving in Panama about 5:20am. The plane was stinking hot throughout the trip and there was no relief at our various transit stops. Miami was very humid and Panama felt like New Guinea. The plane didn't seem to be air-conditioned and smokers were given free rein to poison the rest of the travellers. Service left a lot to be desired.

We left Panama at 6:20am and flew over the entrance to the famous (or infamous depending on the period in question) Panama Canal where at least 20 ships were waiting to cross oceans. The trip south to Quito was uneventful and I enjoyed a very nice breakfast of omelette, Danish pastry, bread roll and compote of fruit at 6:45 the next morning. Quito has the most magical of settings, perched 9,000′ (2,743m) up in the Andes on the edge of a plateau with an escarpment on one side and the eroded 16,000′ (4,877m) volcano Pinchincha on the other. The air temperature was an invigorating 44°F (7°C). We had a good view of Cotopaxi Volcano in the distance, its snow-covered cone rising over 19,000′ (5,791m). Quito's population is less than half a million and the people have a reputation for friendliness. Around 40% of the people are Indian, 40% are Mestizo and 10% European. The remainder are mainly of African descent and Asians. The Indians are characterised by relatively dark skins, short stature and wide-spaced eyes. The airport shops were stocked with an incredible array of craft goods - rugs, blankets, bead necklaces, charms, paintings etc.

We departed Quito at 8:45am and passed right beside Cotopaxi, about 45 miles (70km) south of the capital. It is one of the highest active volcanoes in the world. Also spotted Mt Chimborazo, the highest mountain in Ecuador at 20,697′ (6,310m) and Mt Huascaran, the highest mountain in Peru at 22,126′ (6,746m). The transit lounge was quite cosmopolitan and again there were a lot of shops selling arts and crafts, especially goods displaying pictures of llamas. Most of the houses in Lima, like Quito and other cities on the west coast of South America, are villa style, white-washed and flat-roofed.

From Lima we flew to La Paz, the capital of Bolivia and the highest capital city in the world, with the airport at 13,391′ (4,083m). The city is situated 500m below the airport at the bottom of a steep canyon. This Altiplano region of the Andes is very dry, has little vegetation and is much cooler than the coastal cities. The country itself is twice the size of Spain. The altitude made us light-headed and increased our heart rate. The feeling is not too dissimilar to that of a hangover, so it goes without saying that altitude and alcohol don't mix. The population here is 70% Indian and there is very little English spoken.

The final flight down to Santiago traversed the length of the Atacama Desert, the driest desert in the world. Arica at the northern end rarely ever gets rain and is reputed to be the driest city on Earth. The only places one can find vegetation are in the dry river channels that run from the western slopes of the Andes. Bright green oases dot these river beds, marking the location of small estancias (farms) trying to make a living from orchards, small crops or animal husbandry. Just before Santiago we saw the massive bulk of Aconcagua, the highest mountain in the southern and western hemispheres at 22,841′ (6,962m).

20th January 1971
Santiago

The capital of Chile lies in a valley between the coastal ranges and the Andes and has the same latitude as Sydney. The city itself is dry, dirty and bereft of modern buildings. The people seem very poor and like everywhere else in South America there is very little English spoken. After the humidity of the northern cities it was very pleasant to experience a warm dry atmosphere. It was also nice not to have to struggle with the effects of altitude sickness, as the city is at an altitude of less than 1,000′ (305m). The taxi ride to the city took around 20 minutes and cost 55 escudos (around US$2.50).

We stayed at the Pan Americana Hotel where the tariff was 119 escudos for a double room. The legal exchange rate was 14es per US dollar, but we were able to avail ourselves of the black market rate - around 20es per dollar - through the discrete services of some of the hotel staff. The rooms had private facilities and were very comfortable.

We enjoyed a long twilight that lasted until after 9pm and I recorded the observation that miniskirts were very popular. After breakfast the next morning we took a cab to San Cristobal, a large hill in the middle of the city, where we took a cable car to the summit. The hill is constantly manicured by an army of gardeners, and the view from the top is spectacular. We passed an aviary on the way up but couldn't check it out as access was only from the bottom. The hill also boasts a number of religious statues, including a major one of the Virgin at the top. There is also a swimming pool and a cafe at the summit. We returned to the hotel on foot then caught a cab to take us over the coastal ranges to Valparaiso, Chile's main seaport. The association between Santiago and Valparaiso is much the same as the nexus between Perth and Freemantle. The three hour journey took us through two tunnels, eucalypt and pine forests, cornfields and vineyards for a cost of 450es (US$25.00). Most of the cars we saw were very old and small.

metres = feet X 0.3048

21st January 1971
Valparaiso

Valparaiso is built on the slopes of the coastal range where a deep indentation occurs in the coastline. Thus the city forms a broad amphitheatre facing the sea, similar to Wellington in New Zealand, and the harbour is well protected. At the wharf we located the motorboat to take us out to the Wyandot, the converted American troop carrier that was to be our vehicle for reaching Antarctica. When we reached the ship we located our berths and dumped our luggage before returning to the city for some sightseeing. A US commander took us to Vina Del Mar, the local seaside resort for wealthy Chilenos. It was not very impressive: a few expensive hotels along the waterfront, a couple of nice beaches and a casino. Had a meal there, then returned to the ship.

Arose early the next morning and caught the 8:30am boat to town for another day of sightseeing. Bought the usual postcards and had an empanada for lunch - something like a sweet meat pastie. A crowd had gathered to see El Presidente (Allende) arrive, which he did in a very modest car, not to see us off but for some function or other. Military bands were playing, crowds were cheering - the classical South American pomp and ceremony opera. There were plenty of Argentine tourists but few European or North Americans. The local currency is nearly all note-based, ranging from 0.5es, through 1, 5, 10, 50 to 100es. Coins are not commonly used and the only ones we saw were a large bronze 10¢ and a smaller 5¢. Most of the streets in Valparaiso were narrow and many were planted with palms.

Changing money on the black market was usually carried out in bars or hotels. Hotel porters will change travellers' cheques but bars usually only handle cash. The bars were giving about 22es per dollar whereas the hotels only managed 20. We noticed that the police wear the same uniforms as the army, i.e. khaki. The only difference is that the police have a square green badge with crossed rifles. They carry a gun and a baton. Buses are generally full and doors are left open, with people hanging out. Took a cable car to the top of a hill and obtained a great view of the city and harbour. I could see the famous Chilean training ship, the Esmeralda, a full replica of an old four-masted sailing ship. I could also see four new warships, lying side by side against a wharf. We were fascinated with the streets, which were covered in some kind of yellow tile, often forming quite decorative patterns. We were informed that import duty on cars was 350%, hence the lack of expensive cars on the roads. We returned to the ship around 4:30pm and sailed about an hour and a half later, a bit behind schedule due to the disappearance of one of the crew.

22nd January 1971
Aboard USS Wyandot

The weather and sea were very calm so the prospects were good for a calm crossing to Antarctica. Our stateroom has four bunks, four lockers, a bathroom and a table with three chairs. Owen and I occupied two bunks while a physiologist and a medic occupied the other two. I was pleased to find that the bunks were quite comfortable. However, the locker doors kept opening and slamming shut all night - bloody nuisance. There are about 12 passengers on the ship, destined for Palmer Station, McMurdo and the South Orkneys (us!). The physiologists are planning to study fish and a couple of guys will be studying arthropods (mainly eight-legged mites). One of the passengers is the head of the Antarctic treaty, an agreement among several nations to consider Antarctica a politics-free zone for mutual cooperation in scientific research. The food was excellent and we dined at second shift, which translated to 8:30am, 12:30pm and 6pm. Weather still good. A bit of a swell last night caused one of the scientists to be seasick. The swell is getting heavier as we move further south. There weren't many lifeboats but there were numerous life rafts. The ship is carrying some heavy equipment for McMurdo station, to be offloaded by landing barge. Bird life includes storm petrels and pelicans.

The next day we experienced rainy weather and a slightly stronger swell. Although the ship was now pitching a little there was no rolling action. A couple of the passengers were now sick so I kept taking my Dramamine. Saw some whales and dolphins around 5pm as well as some glimpses of the South American coast around 42°S. The 52 crew are all civilians, and all male. Poor Abby was given very little freedom to move around the ship as it was felt her female status might be unsettling for the crew. Saw some films: The Domino Kid, a Pink Panther cartoon and a film on Basking sharks off Ireland. Sighted some albatrosses the next day and noted that the wind had picked up considerably. My hair has now gone berserk and is all over the place. Ship is doing 14 knots (26kph) and covering about 5° latitude per day.

Sunday morning the 24th and we are now at 48°S latitude. The weather is poor again so spent most of my time reading, eating and sleeping. Took fourth Dramamine but am still feeling alright, amazingly for me. Saw another two films: Mission Impossible and

The Last of the Fast Guns. Weather worsening as we approach the region of Cape Horn. Passed the horn on the 25th and saw the entrance to the Straits of Magellan. Very rugged coastline. The colour of the sea changes from blue to grey and all shades in between, depending on the amount of cloud cover. Our room was very hot today - around 84°F (29°C). One bunk was 90°F (32°C). Fortunately the room cooled to 75°F (24°C) by Tuesday.

27th January 1971

Palmer Station/Anvers Island/Antarctic Peninsula

Saw the outline of Antarctica the following morning and in my enthusiasm took too many photos showing mostly sea and sky and a thin strip of something in the distance. We gradually approached Arthur Harbour at the southern tip of Anvers Island where the US base, Palmer Station, is established on a rocky promontory. The view was awesome.

At the harbour mouth there was a glacier approximately 200′ (60m) thick. We saw three boats anchored near the wharf and a Coast Guard ice-breaker further away. We arrived at the wharf at 12:30pm. I noticed that the wharf is reinforced with vertical steel palings to protect it from winter ice floes. Two of the boats at the wharf were the scientific research vessels Alpha Helix and Hero. The latter would be our home for the next two months or so. The third boat was an Italian sailing boat carrying out soundings in Deception Island harbour. Palmer Station consists of two main huts. One is three storeys high and houses the dining and lounge areas. The other has all the recreational facilities which include table tennis, snooker and darts. The radio equipment is fantastic. There is a separate room for 'ham' (amateur) radio operators and anyone can make a phone patch home as often as they like. The story goes that the last radio operator only set foot outside the building twice - once when he arrived and once when he departed for home. One can't help wondering why he bothered to come down here at all. There was also a good tape music library and an excellent scientific library. Quite a few ships call in here but mail only gets delivered twice a year, the last delivery being eight months ago. The base has a limit of 25 visitors at any one time.

Meal hours were from 7:30-8am, 12-1pm and 5-6pm. After dinner I went for a walk to the glacier edge and saw a chinstrap penguin near a rookery. As the name implies the chinstrap has a thin black line running around his chin and up his cheeks. The rookery smelled disgusting. Also saw a cormorant skimming across the water. Plenty of skuas and Dominican gulls. The skuas "molest" people but don't often hit. They are like large gulls with dark grey-brown colouring. Blocks were constantly falling off the front of the glacier with a sound like cannon fire and small ice caves were visible near the base of the glacier. Apparently the old Palmer Station, which was run by the British, is located around the other side. It is now abandoned as it was only ever meant to be temporary. A large elephant seal colony lives near there and can be accessed by boat or a three-hour trek across the glacier. The rock around the base is Andean intrusive - dioritic by the look of it, and obviously polished by the movement of a glacier. The rocks were also decorated with an array of beer cans, testament to the total lack of sensitivity of many of the personnel to their unique and fragile surroundings. The weather was calm but starting to deteriorate, with a little snow falling. The place is incredibly quiet. The only sound is the hum of the generator in the background and the periodic cannon fire of the calving glacier. In the distance I could see the mountainous form of the Antarctic Peninsula (Anvers Island is just off the coast of the peninsula) with glaciers coming down to the sea.

Palmer Station has all the comforts of home except for women and nightlife. Nine guys are staying on over here, most of them biologists. These include physiologists studying the ice fish - large fish with disproportionately large heads and no haemoglobin in their cells. How do they survive the extreme temperatures of Antarctic waters? and how do they compensate for the lack of iron in their cells? Other fish here are Palmer perch and Antarctic toothfish. A couple of other scientists are doing temperature studies on penguins on an island in Arthur Harbour. They live in a tent on the island. Among the visitors at Palmer are two National Geographic staff, a photographer, and a writer. They are doing an article on the Antarctic Peninsula (see November 1971 edition) which apparently has never been written up by the National Geographic magazine before. They will be travelling on the Hero with us to the South Orkneys. There were a few sailors from the Wyandot scrounging around the rocks looking for full beer cans. Human seagulls!

Owen and I shared a room in the recreation building. Of the 30 or so scientists now residing at Palmer we four appear to be the only geologists. Temperature today was probably in the 30's (-1°C) which is as warm as it gets down here. Winter temperatures get down to -20°F (-29°C). Last year the winter was quite mild and the temperature didn't go below -13°F (-25°C). The next morning was overcast but little wind and no precipitation. Abby and Phill joined me for a walk up to the second flag on the glacier, about 300 yards (275m) away. Then we organised food for our camp on Signy Island in the South Orkneys. In the afternoon we tried to cross onto Bonaparte point on the flying fox. I sat myself in the saddle and pulled myself across and back; very slow and tiring. The channel below was about 80 yards (70m) across. That night the Wyandot sailed and we completed our preparations for our trip. We collected our USARP (United States Antarctic Research Program) bad weather clothing, our two pyramid tents (bit like Scott's expedition), cooking supplies etc. Played some table tennis that evening and had a crazy conversation with a physiologist who believed we all had a duty to get sterilized before having kids to stop the world becoming grossly over-populated. The sun sank just below the horizon around 9:30pm and stayed there all night. It never gets completely dark in Summer.

28th January 1971
Old Palmer Station

The next day was cloudless and very warm - nearly 40°F (4°C). We finished arranging our goods and then the four of us took a motorised rowboat over to the old British base. We found a British hut and a couple of US huts and all appeared to be occupied at the time. The inside of the British hut was a shambles. The US huts were a little better but not much. Apparently the inmates were there to mark the emergency runway. Saw a couple of solitary Adelie penguins and several elephant seals. All the latter appeared to be females, about 7-8' (2.5m) long. The males are much larger and have the interesting distinction of being one of few mammals to have a bone inside their penis. One of the scientists used to use one as a swizzle stick for stirring cocktails. The penis itself is housed inside the body when not in use!

Also saw several cormorants, terns and a giant petrel, and became the target of an aggressive skua which eventually scored a hit. They are 2-3' (0.6 – 0.9m) across their wingspan and are therefore bigger than the notorious magpies. They are the only birds that live at both the north and south polar regions. The penguins were mostly Adelies which are very small, about 18" (45cm) high. They would climb slopes with flippers then slide down on their fronts. Apparently they all migrate to Tierra Del Fuego in the Winter. We saw lots of moss near Old Palmer Station, red and green and various shades in between. Also a few tufts of grass growing between the rocks.

We watched a British ship, the Endurance, re-surveying Arthur Harbour using helicopters armed with sonar equipment. Eventually we returned to Bonaparte Point where we spent some time lazing around in the sun, if not the heat. My arms and calves were bare and I was wearing just a skivvy and trousers. The water was very clear and looked very inviting, so I put my hand in to see how cold it was. It burned! They say a lot of people who fall in get respiratory shock - it is so cold their breathing stops. Needless to say you wouldn't last more than a few minutes in this water.

The Hero returned around 7pm on Friday the 29th. That day Phill and Abby and I made phone patches to New York via a 'ham' in Mississippi. The weather in New York was 18°F (-8°C) and it was snowing so it was warmer in Antarctica! The next day we went across to the Alpha Helix in the afternoon. It is a biological research vessel and is about the same length as the Hero, about 130' (40m). It was about six years old, with several laboratories and two lounge-rooms. The draught was slightly shallower than the Hero, with the design based on the tuna fishing boats of San Diego. The Hero's design is more along the lines of a trawler. Another difference is that the Hero is made mostly of wood whereas the Alpha Helix has a steel hull. The coast guard ship Westwind left at lunchtime today.

1st February 1971
Gerlache Strait

The next day, the 1st of February, was our last day at Palmer Station. We finished putting our gear on board the Hero and checked out our cabin. It was very small and my bunk was so narrow I had just enough room to turn over. Had my passport stamped "Antarctica" and officially signed. That night we slept aboard the ship so that we could get an early start the next morning. Departure was 8am and the weather was perfect again. There was a slight swell initially but this soon settled down and we were cruising through a dead calm sea. Saw a number of whales, crabeater seals, penguins and terns. The crabeaters are much smaller than elephant seals - we spotted them sleeping on ice floes. Clumps of loose pack ice were scattered around and there were plenty of icebergs. The scenery was awesome. The peninsula (also called Grahame Land) displayed every conceivable alpine feature: glaciers, cirques, knife-edge arretes, ice falls and snow covered mountains. Some of the peaks rose to 4,700' (1,433m) and most rose straight from the water. Generally there wasn't much pack ice.

2nd February 1971

Frae and Bellinghausen, South Shetlands

We were now heading for the Chilean base of Frae, and the Russian base of Bellinghausen, on King George V Island, the largest island in the South Shetlands Group. I spent most of the morning on the observation deck with the others, enjoying the vistas and wildlife. During the afternoon we had a fire drill and boat drill. Around 5:30pm the National Geographic photographer spotted some killer whales but they didn't come close. Next morning we arrived at the anchorage for the neighbouring bases of Frae and Bellinghausen. Owen, the two National Geographic guys and two crew members went off in a small Russian boat to shore. Later we went ashore in an amphibious landing craft. As soon as it hit the beach it kept going on its runners. We were planning to spend a full day here but there was a strong onshore wind blowing which was dangerous for the ship so we only spent eight hours. The two bases are built on flat ground and are separated by a mere 100 yards (90m). The Russians have about three huts, in contrast to the Chileans who live in just one large hut. The Russians were very friendly, down to earth and extroverted, whereas the Chileans were very conservative, diplomatic and reserved, although still very friendly. We each received a souvenir envelope and a decorative spoon from the Russians and some stamps from the Chileans.

About 15 minutes after we arrived, a 50′ (15m) Hawaiian sailing boat, the Awahnee, with 6 crew (3 NZ & 3 US) arrived in the bay and dropped anchor. They were in the process of sailing around the world below latitude 60°. They had already been to Palmer and Deception Island and were planning to go to the neighbouring

Elephant Island (also in the South Shetlands) next. Elephant Island was the location of the famous "camp" where Shackleton was forced to leave most of his party after the Endurance went down. One of the New Zealanders from the Awahnee was a giant, nearly 7′ tall (2.13m). Our Abby was the first US female scientist there for which she received a diploma. The wife of the captain of the Awahnee also received a diploma for being the second US woman to visit there. We had quite a celebration in the Russian hut. Plenty of Chilean wine and shishkabobs, Russian soup, macaroni and ham, bread and butter and Russian sweets.

We eventually left the Russian base, saying farewell to Flynn, the photographer from National Geographic, who was going to stay there for a few weeks before joining us on Signy Island. We moved across to the Chilean base, and had another meal. They served us small hot dogs with all sorts of fillings. Also plenty of grog. They seemed to be more organised than the Russian base. Whereas the Russians were entirely scientific, the Chilean base was run by the Air Force and the focus was on meteorological work. A couple of palaeobotanists were staying there for a few days. They had been working on Jurassic and Cretaceous sediments from Livingston Island. Total personnel at Frae is about 60, compared to just 13 at Bellinghausen. The Chileans reckoned the Russians only bathed once a month and that their bath-house was very primitive. When we left, the crew of the Awahnee were about to try out the bath-house but we weren't able to talk to them afterwards to verify the rumours. There was good camaraderie between the Ruskies and Chileans despite the fact that they speak different languages and communicate only poorly. The latter borrow heavy equipment from the Ruskies whenever they needed it. The Russians didn't appear to have any mail system at this base so we posted letters from the Chilean base. Apparently the Chileans are building a runway which should be open to aircraft in about 12 months' time.

4th February 1971
South Orkneys

I spent the austral Summer studying the geology of the South Orkney Islands on the margin of West Antartica, as part of a larger study of the structure and tectonics of the Scotia Arc.

Lousy day. The ship was rolling and pitching which made me seasick. Spent most of the day on my bunk but went out on deck occasionally to see the sights. Saw a 7 mile long iceberg with a flat top - probably part of the Weddell ice-shelf that had broken off. The next day the sea was even worse and a strong wind was blowing. I spent all day on my back. The others weren't too good either. We reached Signy Island in the South Orkneys around 4:30pm. The island is shaped like a triangle and there is a British base at the foot of a glacier on the eastern side of the island. Lots of icebergs offshore. We didn't anchor in the bay because the British ship Biscoe was there. The temperature was 30°F (-1°C) and it was snowing lightly. As there wasn't room in the small boat for all of us I had to remain on the Hero while they went ashore and checked out the base. The base is staffed by 18 personnel, of which 14 are scientists. Most of them are young and working on biological projects. All the personnel here are volunteers in contrast to Palmer where many of the staff are seabees (navy personnel).

After breakfast the next day we departed for Laurie Island and the Argentinian base, Orcadas. That name means Orkneys in Spanish. The anchorage here in Uruguay Cove was much better than at Signy. There are about seven huts at Orcadas including the original one established in 1904 by a Scottish expedition. That was the first permanent settlement in Antarctica and the hut had a wooden floor, four corner posts and walls made of graywacke boulders. The site is on a pebbly beach forming a narrow spit between Scotia Bay and Uruguay Cove. A number of Adelie and gentoo penguins were wandering around the beach which was littered with dead cormorants and the bleached bones of whales. There was a large cross in the centre of the beach which served as a guide to mariners - to the right is safe whereas to the left is dangerous for small craft.

The main hut at Orcadas has accommodation for 21 men, and every room has a fire extinguisher. Needless to say, they take fire very seriously down here. Total complement at the base now is 54, a few more than usual due to the need for building some new huts. One quarter of the permanent residents are radio technicians. The rest are mainly meteorologists, engineers and electricians. The base is controlled by the navy so there is a heavy emphasis on protocol and formality. Everyone knows his place! There was a small cemetery a small distance from the huts, with seven graves. One body was not recovered and two were sent home to be buried. The earliest death was in 1903, the last in 1959. Apparently the last two were lost on the glacier. The base rarely gets visitors, and only one of the residents spoke any English. Most of them were short, with moustaches!

The last published geological work on Laurie Island was done in 1904 by the Scottish expedition. Most of the rock around the base was graywacke, a dirty coloured sandstone. Our main vehicle for getting around the island foreshores for geological work was the zodiac, a five-seater inflatable with an outboard motor. With one crewman we made six landings around Sheila Cove and generally to the north-west region of the island.

One of those landings was to a huge penguin rookery that stank. It was swimming in pink-red penguin excrement, the colour coming from the shrimp-like krill that is their staple diet. We made dip and strike measurements of bedding and cleavage, and noted current bedding, flame structures and loaded ripple marks. All graywackes and shales were tightly folded into asymmetric folds.

Saw some Weddell seals and some small elephant seals; also cape pigeons, storm petrels and cormorants. Some of the cormorants and cape pigeons regurgitated at us when we came too close. We spotted about a dozen fur seals and numerous colonies of chinstrap penguins. At night on the 7th, Owen gave a seminar to the crew on what we were trying to achieve on this trip. They seemed to appreciate it.

The next day was beautiful, 35.6°F (2°C) but one of our landings was a bit 'hairy'. A big swell poured water in the boat and thrust us against the rocks. We all got our feet wet and the spark plugs played up a bit but we were able to keep going. Our usual routine was for Owen to jump out first, then stabilise the boat while the rest of us piled out. If the water was rough the crewman would take the boat out again, otherwise he would tether it to the rocks. As we were studying the structure of the rocks we always looked for shale rather than sandstone as the more ductile shale showed deformation more clearly. We carried two walkie-talkies in the zodiac as well as a spare motor, fuel, oars and life preservers. We also had a survival kit. Normally we started work at 8:30am and knocked off around 4:30 with 10-15 minutes for lunch.

9th February 1971

Clayton (National Geographic writer) went with the others this morning to have a look at the work we were doing. After today we will have mapped most of the northern part of the island. The others saw around 200 fur seals. Apparently the big fur seal colony is on Powell Island. The adults are virtually black whereas the young ones are brownish. Some of the penguins on shore had deep wounds on them, possibly inflicted by leopard seals which are close to the top of the food chain in this part of the world. Weddell and crab-eater seals are not very aggressive whereas fur seals can be quite testy and possibly kill penguins. The coastline of Laurie Island is made up of sheer cliffs, rocky promontories, glacier fronts and tiny pebble beaches. Plenty of bergshrunds at the rear of glaciers. This is the region where moving ice separates from the snowfield, leaving a large crevasse.

The crew told us that some time back one of their number dived into the water and swam to an iceberg about forty feet (12m) away, climbed onto it, then swam back to the ship! This probably took place at Palmer. When Flynn went scuba diving a few days ago a number of seals came up to him and prodded him with their noses. They showed no fear at all, presumably because they had never been hunted. Fur seals are the only seals that are protected. A good fur seal coat is worth around $1000 in New York.

10th February 1971

East coast of Laurie Island

Weather fine again, with the temperature somewhere between 30°F and 40°F (-1°C – 4°C). Saw a lot of wildlife, including a bit of action when a leopard seal grabbed a penguin and whipped it back and forwards until it was turned inside out. On land, seals and penguins are perfect neighbours, but in the sea they become hunter and prey. Saw a lot of fur seals with pups and one huge elephant seal 10-12′ (3-3.5m) long.

We walked across a couple of rookeries, the ground oozing with pink excrement. The stench was incredible. I should point out for the environmentalists that we were sometimes obliged to walk through these colonies to get to outcrops of rock. Some of the rookeries are really huge with thousands upon thousands of birds. One rookery that we estimated to be half a mile long was solidly packed with birds. The chicks are only 6-9″ (15-20cm) high and covered in grey fluff. In contrast to the penguins, seals usually just lie on the beach like sun-worshippers.

Our zodiac was slowly deflating so we headed back to the Hero early. The batteries in our walkie-talkies were low and the Hero couldn't hear us so it would have been curtains if our zodiac had sunk. That night Owen, Aaron (the radio operator), Phill and I went ashore to the Argentine base for dinner. Aaron was wearing thermal underwear, jeans and a parka – his best dinner outfit. I was the best-dressed person there in my flared jeans, brown skivvy and checked shirt! The Argentine commander, the radio engineer and the doctor were also dressed casually. The Awahnee had arrived in Scotia Bay just before the Hero returned there so they dined with us. The cook did an excellent job at such short notice. We had an interesting and tasty, but unidentifiable entree, wiener schnitzel with grilled cheese and tomato, an apple which we ate with a knife and fork and two cups of coffee. As soon as we finished at the table we retired to the lounge and drank whisky. Apparently they had no beer, just spirits and wine. Chatted to the New Zealand guys; they will leave the Awahnee when it reaches NZ in around 40 or 50 days.

11th February 1971

Saw a large rounded block of granite on a beach, which must have been ice-rafted as there is no granite on Laurie Island. Noticed that Adelie penguins have completely black heads whereas chinstraps are only black above their eyes and on the back of the head. Adelies also have red and black beaks in contrast to chinstraps which have all black beaks. Adelies have a small white circle around their eyes. They tend to be braver than chinstraps and stand their ground when people approach. While on the subject of wildlife, we also came across a few Weddell seals although they weren't nearly as common as elephant seals. The Weddells are much lighter in colour and are covered in large spots. On the beaches we often spotted sheathbills, an ugly scavenging bird similar to a gull but pink around the eyes. They steal penguin eggs and feed on carcases of dead penguins and other birds. Saw several giant petrels, called boobies. They can only take off after a good run over the water first. We didn't see any emperor penguins but there is supposed to be one at the Argentinian base called Almirante Brown in Gerlache Strait.

Returned about 2:30pm as a fog was closing in. The Awahnee came into the Bay a little later and they came on board for dinner. Their boat was 53' (16m) long and 11.5' (3.5m) wide (beam) made of ferro-cement over a piping and wire frame. The Kiwis told us that the Russians only have washbasins and a big water heater in their bathhouse - no showers or bath!

The outline of the coast was largely a result of Pliocene glaciation and older Triassic joints. The joints set up zones of weakness which glaciation and wave action worked on. Hence all the small peninsulas around the island are parallel to these joints (oriented NNW-SSE). Had a look at Graptolite Island - a bit of a misnomer as there obviously aren't any graptolites (fossil organisms) present as the rocks are late Palaeozoic and the graptolites became extinct well before then.

12th February 1971
Taxidermy!

Rain, wind and so no work! Owen and I ventured out about 5pm but we only stayed out for about 30 minutes. In that time we managed to walk through two penguin rookeries. One of my outer gloves fell into a pool of excrement! That evening we went across to the Orcadas to learn how to stuff penguins (taxidermy). The crew's quarters were very rough. They eat on benches, off rough wooden tables. Considerable difference in status between the Argentinian officers and ranks. The men sleep in dormitories whereas the officers have their own rooms. The men were very friendly, happy people; not spoiled by the artificiality and superficiality of the more well-to-do society. They wanted our signatures on their Hero postcards.

The next day was too windy to work from the zodiac so we rested up again. Played scrabble and "Oh Hell" with Clayton, Owen, Phill and Abbey. Oh Hell is like Bridge but everyone makes just one bid, which he or she has to try and achieve.

The following day was still windy so just Owen, Phill and myself went out, working along the Mossman Peninsula in Scotia Bay. We climbed halfway up a steep mountain near the end of the peninsula. Eventually we came down a 35° scree slope, the result of extensive frost shattering. We then threw boulders down a very steep ice slope, about 800′ (250m) long and inclined about 65°. At the end of the slope the boulders hurtled into the water with amazing speed. Awesome!

Wonderful view of Scotia Bay from the mountain - lots of icebergs floating around. Some are bright green, possibly from being submerged for a long time before being overturned. Returned to the Hero around 5pm. Spent the evening playing Scrabble and Oh Hell, then editing Clayton's National Geographic report on our geological activities (see report in November 1971 edition).

15th February 1971
Seals and Liverworts

We worked the west coast in good weather, observing numerous crabeater and fur seals in the afternoon. The fur seals are the only seals to walk on their hind flippers. If elephant seals tried that they would break their flippers! We also saw a bit of plant life, mainly liverworts, mosses and pale green grass. The liverworts were orange and black with small caps on their surfaces.

Most of Laurie Island is composed of steep cliffs, ice and snow covered, with several fairly tall peaks. Many places show plateau tops suggesting an uplift in the land in the recent past. Small icecaps and piedmont glaciers are scattered over the island, making land journeys untenable. The cliffs are characterised by steep scree slopes – around 35-45°. Did some 'hairy' rock-climbing today. Owen romps up the cliffs; guess who trails behind! Over the next couple of days we continued working around the island, getting wet and cold but managing to complete quite a bit of work.

18th February 1971

Murray Rocks and Graptolite Island

Owen, Clayton and I went around to Murray Rocks (where there was a cormorant rookery) and Graptolite Island while Phill and Abbey worked around Orcadas. Noticed some red and green snow algae. We saw a couple of macaroni penguins near Graptolite Island. They have yellow-orange whiskers and are very rare in the South Orkneys. Also saw a number of elephant seals and gentoo penguins, and a couple of leopard seals on an iceberg. One penguin jumped into a purple-brown pool of excrement and regurgitated krill. He swam through it then blinked a few times to clear his eyes. The penguins seemed to be gathering into groups in preparation for the long annual swim to Tierra Del Fuego where they will wait out the Winter. Last year penguins left Heard Island in January.

The island is predominantly sandstone with lesser shale, strongly jointed in a NW-SE direction, parallel to the peninsulas. Glacier calving was quite active, with a noticeable time lapse between seeing the ice fall into the sea and actually hearing it.

19th February 1971
Frederickson and Powell Islands

Sailed for Frederickson Island today, arriving there about 9:30am. It is about 2 miles (3km) long and generally flatter than Laurie. We travelled around it in the zodiac, observing that there was much more shale and some conglomerate. The scree slopes are more shaley and the structure appeared to be chaotic. Prior to our arrival only one or two landings had ever been made on Frederickson Island. We made about eight landings around the island. We also stopped off at the southern point of Powell Island. Powell Island is much larger than Frederickson and is largely snow and ice covered. The southern point is very flat and we were able to stride around instead of having to climb. The whole island appeared to be composed of conglomerates. One of the crewmen found a Japanese fishing float on one of the beaches. Weather was showery and cold. At the end of the day we returned to the Hero and 'set sail' for Signy Island where we planned to camp for two weeks.

20th February 1971
Signy Island Camp

After breakfast we started taking our gear ashore to our campsite. The site is just across the bay from Signy Base, on a patch of level ground, quite sheltered and only a couple of hundred yards from a freshwater lake. It is also close to a pebble beach where a couple of bull elephant seals and numerous cows reside. The beach was our toilet! The rocks around here are marbles and garnet schists, quite a change from the un-metamorphosed sandstones and shales of Laurie Island. Moss is very abundant here, along with some grasses and liverworts.

After lunch on the Hero we set up the three pyramid-style tents. They are called Arctic Pyramids (Black & Co., $350), are made of Ventyl, a tough wind-resistant material, and have double walls, providing about 9 square feet (1 square metre) of floor space. They have four poles and a ventilation sock near the apex. There was sufficient space between the walls to put our work gear and boots. Owen and I shared one, Abbey and Phill lived in the second, and the third served as our mess tent. It was great to be camping again.

Passed a whale graveyard about 300-400 yards (300m) from camp. Lots of whale ribs and vertebrae. Skuas attacked us here and Owen got hit twice. We always let him lead, that way he copped the attacks! The island is a lot flatter than Laurie and slightly swampy in places. We observed that the Bransfield, a large British cargo and personnel ship, was also in the bay (Borge Bay). A number of British scientists were leaving after a stay of 2.5 years. Normally an Antarctic stint for them is two years, but they have to wait for ships to become available.

The temperature at 6pm tonight was 29°F (-2°C) and there was a light wind blowing. Seals burping on the beach, the radio playing in the tent and the kerosene lamp burning softly provided a pleasant backdrop to our situation. We had air mattresses so were very comfortable.

21st February 1971
Around camp

Weather was foul; rain and strong winds with a temperature around 33°F (0.5°C). Phill cooked breakfast: oats and scrambled eggs, fruit juice and hot chocolate. We spent the rest of the morning organising the camp. Buried a couple of boxes of frozen meat in the snow, some distance away as there was little snow around the coastline. A large group of UK guys from the Bransfield and Signy Station came past on their way to North Point. Most were just going for a hike but Clayton and two geologists stayed here and talked for a while.

I have been assigned to cook dinners for a week, Phill is on breakfast duty and Abby does all the washing-up. Owen helps whenever he is not busy. We decided to set our clocks forward an hour as it gets light at 4am and the sun sets at 7pm. So we are now in line with Signy Station time and two hours ahead of Palmer Station time.

The seals on our beach are mainly elephant seals with one crab-eater, although a couple of fur seals live about 50 yards (45m) away. Cooked some lamb steaks for dinner tonight, along with mashed potatoes and tinned asparagus. After dinner there was a queue at

John's beach (the 'John'). The wind dropped in the late afternoon and with the temperature hovering around 35°F (2°C) it was very pleasant. The temperature of the water was also 35°F. An elephant seal tried to get in the zodiac then satisfied itself by lying beside it. The seals become active in the late afternoon and evening. This afternoon I chipped garnets out of some boulders of schist. Most were around 1/4 to 1/3 of an inch in size. Our beach is called Waterpipe beach and it is not far from a small piedmont glacier (Orwell Glacier) which we have to cross if we want to visit the British base. At the foot of the glacier is an area of marshy ground riddled with elephant seals, appropriately named Elephant Flats. In the other direction is a small lake with the remains of a hut at one end of it. An old water pipe, now in pieces, extends from the lake to the beach.

22nd February 1971
Field work commences

Our first working day on Signy was a structural reconnaissance of the island. All four of us walked around to get an idea of what the folding was doing. Rained on and off all day but thankfully there was little wind. Temperature peaked at 32°F (0°C). Moss, 4-5" (10-12cm) thick formed a thick carpet all over the ground. There was also quite a bit of grass, with green cup-shaped flowers. Skuas constantly attacked us. Saw lots of Antarctic terns and giant petrels. We climbed Jane Peak, at 600' (180m) the second highest peak on the island, then walked north to Stygian Cove and west to Port Jebsen. Near the latter we entered a British hut and made ourselves some tea. There are three huts scattered around the island, apart from the main base, and they are just large enough for a couple of bunks and a stove, and a bit of cupboard space. They were built as an overnight shelter or as a short-term base for scientific research. Just after lunch a damp thick mist descended on the island so we had to find our way back using compasses and a map. The island has a sort of eeriness about it – similar to the Scottish Highlands I imagine.

We returned to camp about 5pm very weary after walking all day. I cooked up some fried rice: prawns, onions, rice, soy sauce, curry and chili powder! The rice turned out terrific despite not being able to see the pot properly (poor lighting) – just kept adding water when I thought it needed it. We then had fruit salad for sweets. Afterwards Abby cleaned up while I cooked some more prawns for tomorrow night's prawn cocktail, then heated some water for a wash.

23rd February 1971

Foul day again. Snow during the night spread a thin blanket over everything, which heightened the eeriness of the place even more. Temperature only reached 22°F (-5°C) and a light wind added to the discomfort. During the night two Pommies came ashore when their outboard motor conked-out. Owen gave them a nip of Scotch and tried to call the UK base to advise them of their predicament. However, we couldn't raise their frequency so I lent them my torch and they made their way back over the glacier. A search party was just setting out from the base when they saw the torch light on the glacier.

After the Pommies left we were invaded by seals that had decided that our tent was an ideal wind-break to sleep beside. So with great burping and other choice noises, one settled down right beside where I was sleeping. This was well past midnight and both of us were thoroughly awake. So I gave the side of the tent a great kick and yelled at the intruder. This was a successful manoeuvre but proved to be only a temporary respite. I fell asleep shortly afterwards but Owen was disturbed for quite a while.

Getting up next morning was quite a task. Had another day of arduous walking and my legs were quite stiff. We walked down to Gourlay Peninsula, or at least the bluff overlooking the peninsula. The only penguin rookeries on this island are at North Point and on the Guerlay Peninsula in the south. Our lunch (cheese, liverwurst, crackers and chocolate) lasted about 10 minutes because it was bitterly cold. This afternoon we started taking measurements of the folds.

On the way back we stopped at the UK base for a chat and some coffee. Their base is staffed all year round and consists of a large steel hut and a slightly smaller wooden hut. The Poms were very friendly – all biologists of some sort or another (except the cook and a few other support staff). The engineer at the base was banding cape pigeon chicks and adults. He averages 150 per day and gets covered in regurgitate in the process. The birds have been picked up in NZ with these bands on them. Similarly, storm petrels have been picked up in New York after banding at Palmer Station. We borrowed some sheepskins from the Poms to put between our lilos and sleeping bags so that condensation would occur at the lilo/sheepskin interface and not at the sleeping bag/lilo boundary. Back at camp, a female elephant seal tried to get close to our tents again so Owen shooed it away. It was possibly the same one that visited us last night.

Owen called the Hero on the radio that night as part of our regular contacts. They had had 20-30 knot winds (37-56kph) and massive waves, and nearly everyone on board had been sea-sick.

24th February 1971

Terrible morning – snow flurries, rain, strong winds and a temperature around 28°F (-2°C). The weather improved during the day but remained patchy. Owen and I worked in the Gourlay Peninsula area while Phill and Abby worked north of the camp. We walked along a snow slope at the base of a bluff, appropriately named the Kaiber Pass, to get down to the peninsula. Found a couple of chinstrap colonies and two of the Pommie scientists. They were studying penguin guano for a Masters degree.

Numerous skuas attacked us today – rather nerve-racking. We walked over the Orwell glacier for the second time. Back at the base we found a seal right inside the zodiac. Phill and Abby had come back early because Phill had injured his knee somehow. Abby cooked up some ham and pineapple for dinner. Had strawberries for dessert. Phill is going to put strawberries into hot cakes tomorrow – what a waste of good strawbs!

25th February 1971

Arose as usual at 7:15. Temperature was 24°F (-4°C), overcast but no wind. I felt like a lump of lead as we walked along the Kaiber Pass again. Getting around the base of some cliffs was a little hairy (eg. a strong jump onto a precarious rock face overhanging the icy water). Our traverse along the base of Rusty Bluff was quite tricky. One of the UK guys didn't realise it was possible. Spent a lot of time measuring bedding/cleavage intersections, plunge or pitch and trend of fold axes, attitude of fold axial planes, joints and joint intersections etc. Also noted the vergence of the folds (direction of overturning). We found the grave of a 23-year-old Englishman in a penguin rookery. My comment just before we saw the grave, as we were walking through inches deep penguin shit, was "to slip here would be a fate worse than death!" As we walked back to camp the only sound audible was the crunching sound of our boots over frozen snow. Crossed the glacier again and noted some narrow (less than 3' or 1m) but deep crevasses. Back at base I cooked up some pea soup, then steak and carrots followed by pineapple and strawberries.

The next day the weather was terrible – driving rain with temperatures around 30°F (-1°C). As a result we started work a little later than usual, mainly working around the camp, John's Beach, Elephant Flats and the lake. Came back to camp for lunch and collected some specimens in the afternoon. For dinner I prepared pork chops, peas and apple sauce, followed by more strawberries and pineapple. I had four pork chops and copious quantities of everything else. My clothing today consisted of cotton socks, wool socks, mountaineering boots, cotton undies, thermal undies, skivvy, shirt, trousers, polo jumper, lined parka, waterproof trousers, outer parka, wool gloves and outer work gloves.

Many of the rocks around here have thousands of garnets in them. They are dark red and probably almandine. Tomorrow night we will leave camp around 4pm and go across to Signy where we will shower, shave, shampoo etc. then dine and possibly spend the night there. It will then be nine days since I last washed my hair – a record for me. I am reading 'Five Smooth Stones' at the moment.

27th February 1971
The British Base at Signy

On returning to camp today we found two seals lying on our food supply and several scattered around the camp. We had lunch in Three Lakes Hut where we met Phill and Abby. Two Poms came over in the skiff in the early afternoon and gave us a lift to the Base. After cleaning ourselves up we drank steadily through the evening and night. After four sherries, two glasses of red wine, one Gaelic coffee, one scotch and one rum-and-coke I felt a little tipsy... We played records, chatted and generally clowned around. Retired to bed about 2am. Phill and Abby retired early, then Owen a bit after 1am. Just before I hit the sack they played Australia by Spike Milligan. We ended up staying the night there, and lunch the next day as well.

Despite the place being quite comfortably furnished they had a very primitive can toilet and equally primitive washing and drying facilities. The drier is a clothesline suspended over a heater, and the washing machine is the ringer type. When I washed some clothes there, my thermal underwear top, which caters for a 38-40″ chest, shrank to a shoulder span of 8″.

28th February 1971

Returned to camp around 2:30pm and found the place had been taken over by the seals again. The temperature this afternoon was just 22°F (-5°C), resulting in the tent straps freezing and a thin layer of ice forming on the tent walls. Owen and I packed rucksacks and set off for Foca Hut where we intended spending a few days to work the western side of the island. We had to cross the island with packs weighing between 60-70lbs (around 30kgs). We tried to cross an ice-field but gave up half way up the slope and followed a longer but safer route. Coming back down the slope Owen used his rucksack frame as a sled. Mine had no frame so I had to slide down on my backside. With the wind and fine snow whipping us, a heavy pack, cold temperatures, runny nose and watering eyes, it was quite a walk. The overcast conditions made the place look particularly desolate.

The hut was draughty and cold but was adequate for our needs. We dined on two steaks and half a can of peas each, followed by a mug of tea, biscuits and tinned fruit.

1st March 1971

We worked from 8am to 6pm the next day, enjoying beautiful sunshine until late afternoon when a dense mist covered everything. That evening I paid a visit to the inter-tidal zone and was rewarded with a natural bidet. Timing the waves is very important when one's pants are down.

The next day we set off in light rain for the SW corner of the island. After we lost each other twice we decided to work in Port Jebsen instead. We were feeling tired after yesterday so we knocked off about 3:30pm. Neither of us had much enthusiasm to continue. Skuas were constantly swooping on us and giant petrel chicks would spit at us if we came too close.

The next day was probably the best we have experienced on Signy. Wispy cloud and a faint breeze, temperatures in the 30s (Fahrenheit!). I took some specimens back to camp and brought some supplies back to Foca Hut to last the next three days. The hut is about 11′ (3.5m) by 6′ (2m) and houses a double bunk and a single bed, complete with air mattresses and down sleeping bags. There was also a heater, a wooden table and a couple of primuses for cooking. Boxes of canned food occupied the space beneath our beds. Water is supplied by a meltwater stream a few metres from the hut. The other day, when the temperature was 22°F (-5°C), quite a few of the smaller streams froze over. Our night time routine, once we finished our field work, was to light the primuses, make a pot of tea and put on our dinner (usually steak and a can of veges). Then I washed up and sorted out the specimens while Owen plotted up the data. We generally hit the sack between 10:30 and 11pm. The hut was called Juniper Hall and was apparently built in 1960. The three huts on Signy are all linked to the British base by telephone.

At midday I walked over to the camp via the ice cap around the back of Jane Peak. At one point the ice kept collapsing beneath me and I would sink up to my shins in icy water. The last section of the ice cap was very steep so I sat on the pack frame and tobogganed down.

While I was organising more supplies for the hut, Phill and Abby returned from their field work. We chattered for a while before I returned to the hut with a load of steaks and other supplies. Owen told me of his close call when he slashed his wrist on a rock. Fortunately it missed the artery. He washed it in seawater, put a bandaid on it, then wrapped it in a specimen bag and kept on working. It must have been a bad time for a while there. That night I set my trousers on fire while I was drying them over the primus. I sewed around the burnt patches to prevent them becoming ragged. What with bleaching and burning, they are developing quite a lot of character.

4th March 1971

Had to work in a fine mist all day. At 3:30pm we returned from Thulla Point to find four chaps from the British base having a cuppa. They had been ringing giant petrel chicks, managing to get through around 500 per day, and were covered in regurgitate. One of the Poms stated that the minimum temperature ever recorded on Signy was -29°F (-34°C). Each Winter it often reached -22°F (-30°C).

The next morning I set off to collect specimens while Owen geologised around the hut area. The weather was extremely changeable all day. One moment it was brilliant sunshine and the next I was enveloped in mist. Saw an elephant seal at Foca Point lying in indescribable filth. They lie in their own excrement for days. Pools of black liquid abound, and many of the flat-lying rocks are covered with crap. The stench is really obnoxious. Later, the four Poms came around again.

We left the hut the next morning, wearing snow goggles on the ice cap as the reflection off the ice was very strong. Halfway down the toboggan strip my pack frame turned side-on and I fell off. I slid the rest of the way down on my backside, connecting with a large rock at the bottom of the slope. Fortunately I wasn't travelling very fast at that stage so there was no damage. Although the frame was useful for tobogganing, I came to the conclusion that a rucksack is better than a pack frame for mountaineering work as the frame interferes with one's balance.

On getting back to camp Owen had a look at the radio log and discovered that the vessel Staten Island had hit a rock and was heading back to Australia. This meant that the Hero would have to stay at Palmer Station a little longer, and we would have to stay here a bit longer. Our basic food supply was okay but we had run out of chocolate, drinking chocolate and other luxuries.

We went over to the British base around 5pm, had a bath and beard trim, then dined on soup, fish, roast lamb with baked potatoes, Brussel sprouts, cauliflower, peas and ham, followed by baked custard and some kind of sweet sauce. I pigged out on the food but drank only modest quantities of alcohol. The Poms showed some slides of the base in Winter. The whole of Borge Bay becomes frozen over to a depth of about 6″ (15cm). For entertainment they cut a hole in the ice and go diving, attached by a rope to someone on the surface. Five tugs on the line means distress, four means coming up, three is for going down, two for stopping and one for indicating all is well. They even dive at night under the ice. Last year seven out of ten dived, so all are encouraged to do it. Sometimes they travel out to the sea ice by skidoo but more often than not they ski out in their wet suits. The depth of the bay is 45′ (14m) so there is little risk of the 'Bends' or other problems associated with deep diving.

They had a fancy dress party a year or two ago and showed guys dressed as women. One slide depicting the morning after showed a half-empty bottle of scotch and a bottle of Andrews liver salts (a laxative). Everyone was in stitches for several minutes.

They collect penguin eggs in the Spring and make six-egg omelettes – the whites firm up but remain colourless. Apparently the amino acid sequences in the albumin are different to hens' eggs. Penguin eggs are at least twice the size of hens eggs. The gentoos are the first to lay eggs, then the Adelies and finally the chinstraps. They also showed a slide of a chinstrap with brown colouring instead of the usual black. They find one or two of those every year.

The work of the scientists includes the study of microclimates, physiology of the ice fish and penguins, and soil and moss studies. They also attach metal or plastic rings to birds to study their migration patterns. The inscriptions on the metal rings read "Inform British Museum London SW7 5039048". Wilson's petrels have been picked up in New England after being ringed in the South Orkneys. Personnel can leave after one year but most stay for the allotted two or three year stint.

Coming back over to the camp in the evening the weather was really foul, causing us to get thoroughly soaked.

8th March 1971

We woke to find it snowing heavily with a strong wind blowing. It didn't improve so we ended up having a relaxing day. The temperature at midday was around 28°F (-2.2°C). The water temperature was 35°F (1.7°C – try swimming in that!). The next day was similar, with horizontal snow making life unpleasant. We set off about 9:30am, returning at lunchtime. We set aside some whale bones, which we collected from the whale graveyard (a beach where a couple of whales had died), for shipment back to the US. Out on the hill tops the wind was about 25 knots (46kph). During the night the seals invaded us and kept up a steady burping harangue. I finally arose around 6am and made them bugger off. As soon as I had settled myself back in my sleeping bag they came back again. However, they made less noise this time so I left them alone.

The weather improved a bit the next day, although we continued to get snow flurries. The whole island was covered with about 3″ (8cm) of snow, with the temperature hovering around 26°F (-3.3°C). We walked to the summit of Robyn Peak (700′ or 200m high), then around North Point and back. A group of gentoo penguins arrived at our beach in the late afternoon, probably *en route* to Tierra Del Fuego. That night Phill and Abby cooked hamburgers (again).

11th March 1971

After arising at our usual hour of 7:30am Owen and I walked over the glacier to the British base, spotting a crew mate motoring over to our beach as we were descending the glacier. He returned shortly afterwards and gave us a lift to Moe Island in the dinghy. We had to turn back as the sea was getting rough, so we did some work in Paal Harbour and then returned to the base around 3pm. While I was there I bought two first day covers of British Antarctic stamps for $12; one of these was sold in 2003 for nearly $200!).

The next day was uneventful, although we found out that the Hero had left Palmer at 4am and was expected at Signy around the 16th. I collected some more garnets, some as large as 1″ (2.5cm) in length, and chased some elephant seals further down the beach so they wouldn't disturb our sleep. The chasing process involves throwing rocks at them (gently!). They rear up with their mouths gaping and begin to retreat, always facing their adversary. When they get several yards from you they turn around and go somewhat faster.

13th March 1971

After a late start Owen and I did some work between Elephant Flats and the British Base, then lunched adjacent to the whale graveyard. Around 2:30pm we went over to the base for our weekly shower and grog session. Phill and Abby joined us about 5pm. The Poms told us the names of the only two angiosperms in Antarctica, a grass called Deschampsia Antarctica, and a pearlwort called Colobanthus Quitensis. The latter is a small gregarious plant with triangular shaped fleshy leaves. They also told us that the sea around the base (Borge Bay) froze to a depth of 21″ (53cm) last year. It can take a man's weight when it is 3″ (8cm) thick. Sometimes pack ice comes in from the Weddell Sea pack and mixes with the frozen sea ice, creating a very hummocky surface.

Found a book on penguins that listed all the species:

Emperor, king, Peruvian, fairy, rock hopper, royal, Magellanic, Galapagos, Adelie, gentoo, Snares Island, macaroni, white flippered, chinstrap, Fiordland, erect crested, black footed, yellow eyed, and Chatham Is. blue.

After breakfast the next morning Owen went off with some of the British Antarctic Survey (BAS) blokes to look around Cummings Cove while Phill, Abby and I explored the Gourlay Peninsula. I left them at the bottom of Rusty Bluff and walked up to take some readings. A snow storm came up so I headed back to the base, reaching there about 2:30pm. The others returned from their bird-watching shortly afterwards. We sat around until 6pm when we were collected and taken back to our camp in the boat. In the process of disembarking the skiff I submerged one of my legs up to the knee, which meant removing one boot, two pairs of socks, long johns and trousers. Snowed during the evening (temperature around 25°F [-3.8°C]) then rained in the early morning.

15th March 1971

Went off to collect some amphibolite samples for geochronological purposes (determining the age of rocks). Back at the tent in the afternoon, Owen was able to find out from South Pole Station that the Hero was expecting to reach us at 9pm tomorrow. The ship had encountered a lot of ice on the way to Signy and had been forced to detour to Gibbs Island and then head northwest.

16th March 1971

Terrible day again. Very variable, periods of heavy snow and wind alternating with calm, dry periods. We all slept in and had a lazy morning. After lunch we all went for some exercise on different errands; Owen went to the base, Abby and Phill went north and I went looking for amphibolite for age dating. During the afternoon when I had retired to my tent I tried to light the primus to heat the tent. Things didn't quite work out that way and I had a few seriously bad moments. The primus sent huge flames shooting up to the top of the tent so I picked it up and hurled it through the opening in the tent. Fortunately it passed through both the inner and outer skins and landed harmlessly on the ground outside. Burning kerosene spilled onto the floor of the tent but I managed to put it out without too much trouble. Needless to say the whole little episode gave me quite a scare.

The Hero arrived around 7pm to great applause. I think we were all pleased to be getting off the island after three weeks of close living. Owen was a very good leader, very considerate and good at preventing dissention in camp. What I would miss was the wonderful landscape, the long walks over rugged terrain, crossing the glacier, climbing the peaks and looking at the icebergs drifting in the bay.

$F - 32 = (9/5)C$

17th March 1971

Two zodiacs from the Hero came over to our beach and collected Owen and most of our supplies around nine the next morning. The rest of us retreated to my tent as it was the only one left standing. The weather deteriorated and by 11am a blizzard was blowing. The Hero launched the whaleboat and with one zodiac they collected us and the remainder of our equipment. We finally got on board around 1pm, and probably just in time as the weather really packed in after that. During her cruise to Signy to pick us up the Hero had become stuck in the ice. The ice was flow ice which had built up behind the 16 mile monster berg we came across before. The plight of the Hero had been described in some of the US newspapers. They were going to fly planes in to find a path out of the ice for the ship, but it worked its own way out and continued on to our rendezvous.

We found out that one of the crew had secondary syphilis so everyone on board had to have penicillin injections, and we are supposed to have blood tests when we get back. That will make an interesting claim against Owen's USARP grant! We also found out that the Wyandot was fired upon by the Ecuadorian navy during her return to South America. Not sure why but the captain of the Ecuadorian navy ship was subsequently relieved of his command.

During the night the wind picked up to 35 knots (64kph) with blizzard conditions prevailing for most of the next day. Owen managed to get across to the British base to say our goodbyes. The Poms gave each of us a signed photograph of the base in Winter-time as a memento of our trip.

Next morning Owen and Phill went ashore on Coronation

Island (Cape Hansen) to collect some paleomagnetic samples, returning shortly after lunch completely soaked. The Hero then took off for Laurie Island where the four of us did a bit of work on the west coast. For the last 36 hours, despite the Hero rolling 35 degrees each way, I hadn't been sick. The temperature in the cabin was 69°F (20.5°C) and felt like 100°F (37.7°C). I guess this was because we had acclimatised to 32°F temperatures (0°C) while we were camping.

19th March 1971

Arose at 7:15am then departed in the zodiac for Powell Is. (still part of the South Orkneys). Powell Is. is very similar to Laurie Is. in size and shape, though oriented differently. We went around one side in the morning then returned for lunch and the Hero took us to the other which we visited by zodiac. We made about 14 stops on the island altogether. I got a very wet backside when the zodiac deflated a little and at some stage my feet got wet so by the time we returned to the ship I was quite damp and cold. Finished reading another book during the rest of the day – C.S. Forester's 'Flying Colours'. That's two books in two days.

The next day was reasonable so we worked around parts of Laurie Island. We saw a fin back whale while we were in the zodiac. I read another book during the late afternoon and evening – Alistair Maclean's 'Force 10 from Navarone'.

On Sunday the weather was overcast, foggy and snowing (all day) but very little wind. The surface of the water was very smooth. During the morning I slipped on a rock in the intertidal zone on

Laurie Island and got drenched by a big wave.

Tonight we are going to Orcadas for dinner. Their dinner hour is fairly informal but we expect to eat around 8-9pm. There will be 17 of us going ashore (including 4 geologists, 4 biologists, and 2 USARP personnel). My memory of the evening was drinking endless quantities of Argentinean white wine, champagne and scotch. Food was served in the form of savouries so we all sat around a couple of large tables and helped ourselves to this and that. Abby cut a big cake and made a short speech. Masses of people were roaming around taking pictures all night.

Coming back from the party the next afternoon most of us were in various degrees of inebriation. Despite the zodiac having 5" (13cm) of water in the bottom we managed to get back to the Hero safely. We couldn't work the next day as a storm was approaching and the weather was already quite foul. The wind gradually increased up to 50 knots (93kph) during the day but we were safely anchored in Scotia Bay so were in no danger. Owen decided to leave the South Orkneys as soon as the storm abates. Our work here is finished. I spent the day organising our equipment and packing things in boxes. Tomorrow I am rostered to do duty in the galley, washing up. We all have to assist in the running of the ship.

24th March 1971

Although the weather has cleared there is supposed to be a storm in the Drake Passage so we will wait another day.

25th to 28th March 1971

Crossing the Drake Passage

We endured gale-force winds nearly continuously so most of us were suffering hell. The ship rolled 40° most of the time and sometimes 50°. Became seasick on the third day out even though I was on Dramamine. I spend the days, the hours, the minutes lying prone, doing absolutely nothing. Owen doesn't eat much and rarely attended meals. Abby and Phill were better sailors. The ship leaked a bit and the deck was continually splashed with seawater. Most of the passengers spend most of the days on their bunks and I finished reading 'Von Ryan's Express'. Apparently the ship was being navigated by compass and sextant.

29th March 1971

Still in the Drake Passage but expect to pick up the marine pilot in the Straits of Magellan on Wednesday, nearly six days after leaving Scotia Bay. After we pick up the pilot it will take another ten hours. Even in the lee of the land the ship is heaving considerably. We have been heading into the wind so the pitch is strong. Not much roll though. Now reading Michener's 'Caravans'.

We are now sailing up the SE coast of South America approximately 20 miles (30km) offshore and sailing directly into the wind. If the wind swings to the NW we may be tossing like a cork all the way into Punta Arenas.

We picked up the pilot about 3:30am on the Wednesday. He came out in a glorified landing craft as the proper vessel had been sunk by the Argentine navy about 1.5 years ago, with the loss of several lives. He also had the mail with him so we read that (two letters from Dad) and then retired to get a few more hours sleep. Owen, Abby and Phill are going to sleep at a hotel tonight but I intend to stay on board. I intend to hop across to Tierra Del Fuego tomorrow or the day after.

The land on either side of the Straits is like desert; flat, brown and very dry. The land is like this for about 70 miles (110km) past Punta Arenas apparently. Still haven't seen a tree since we left Valparaiso. Porpoises abound in the Straits. The waters can become exceedingly rough with 80-knot (148kph) winds hurtling down the mountains and being funnelled along the straits. It would be a couple of miles to either shore at the narrowest part of the Strait. The Patagonian shore was illuminated at various points by burning of natural gas. Apparently there are quite a few oil wells in the area, as well as further south in Tierra Del Fuego. Further west where the mountains rise is a wall of grey. There is a cloud build-up in that direction most of the time. Due to the high winds we may not be able to enter the port tonight. The wider parts of the Strait are 16 miles (25km) across. Last night I was told of a German guy who was paddling around the islands in the vicinity of Cape Horn in a kayak. The Chilean navy eventually forced him to abandon the project.

Above: Iceberg from Hero.   Below: Antarctica!

Above: Chinstrap penguin. Below: Elephant seal in the zodiac.

Above: Penguin rookery. Below: Signy Island UK base.

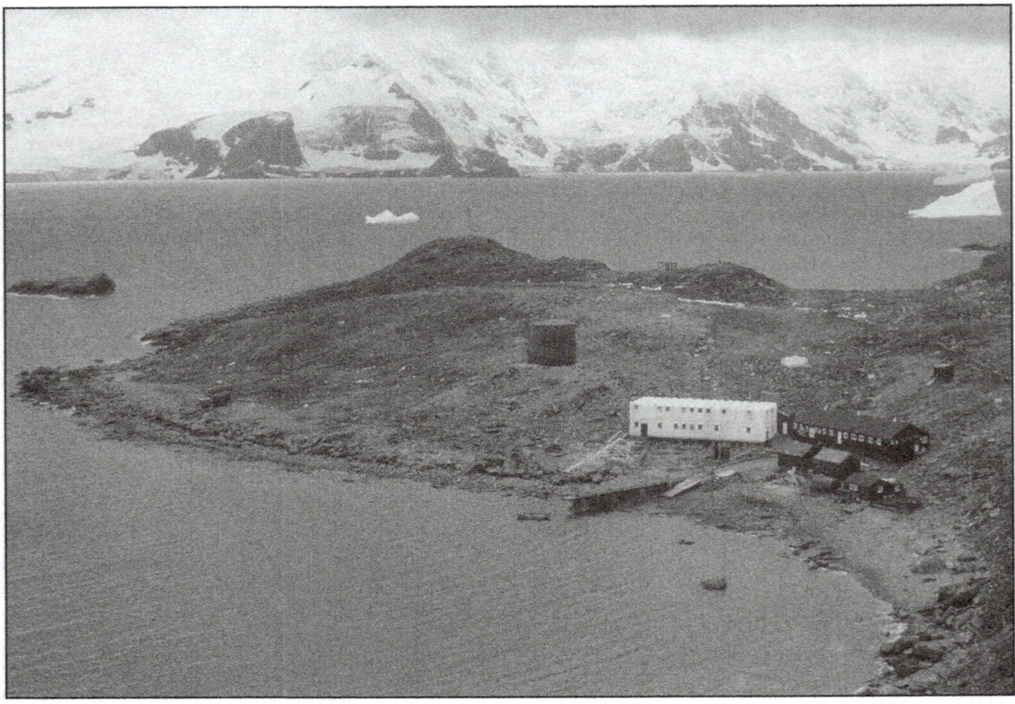

Above: Signy Island camp. Below: The Hero at port.

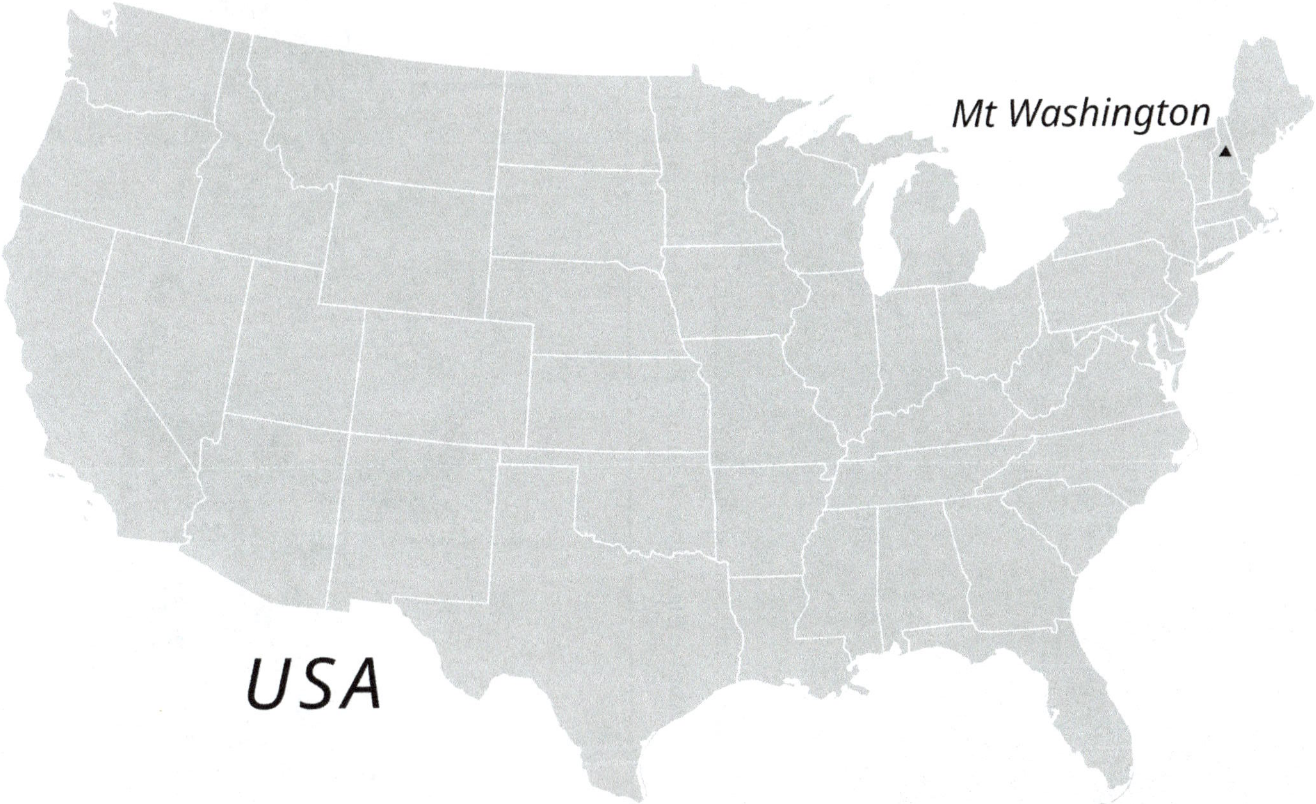
Mt Washington
USA

Adventure to Mt Washington

December 1971

On Sunday 19th December two Lamont-Doherty inmates (Mateus, a Brazilian PhD student, and Kevin, a research fellow) agreed to accompany me on a trip to the Presidential Range of the White Mountains in New Hampshire to try and climb the notorious Mt Washington. The mountain is 6,288′ high (1,917m) and receives frequent high velocity winds as they accelerate up the north-south trending mountain barrier to the summit of Mt Washington. The weather station at the summit has recorded the highest surface wind speed in the world – 372kph on the 12th April 1934. On average, the mountain experiences hurricane force winds every third day and over 100 people have died on the mountain. Wind chill temperatures can reach -120°F (-84°C).

Kevin and I were ready to leave Lamont at 9am but Mateus arrived five hours late on account of getting lost twice. Mateus drove, heading for Albany en route to New Hampshire. We spent the evening in a local pub where the publican kindly put on a free smorgasbord for the drinkers. Later we retired to the car to sleep. The car heater was not working so we were forced to spend the night sleeping in a very cold car parked on the side of the road. Mateus slept in the front seat while Kevin and I stretched out in the back. Mateus had no sleeping bag!

The next day we travelled most of the day and stopped at a pub called the Green Barn near Twin Mountains. Because the place had just opened we were given a free pizza. We managed to get the car heater fixed during the day but then the car developed rear light troubles – called to our attention by a local highway cop. We were not able to get that fixed until the final stages of the trip. Another night spent sleeping in the car, this time with Kevin sleeping in the front.

On Tuesday we attempted to drive up the approach road to Mt Washington but after proceeding about two miles (3km) we were forced to return. Most of the day was spent trying to reverse down the snow and ice-covered road; we dug the car out of the snow five times with only a geology hammer. After a commendable ten point turn the car was able to proceed forwards, which it did under the fearless guidance of Mateus whose philosophy of more speed, less digging, proved the answer. A quick coffee at the Green Barn firmed our resolve and we set up the approach road on foot carrying bivouac equipment on our backs. The six mile hike to the base station at 2,700′ (823m) was completed in about three hours. It snowed lightly most of the way. The caretaker at the station gave us a room for the night and in return we listened to his philosophies on women and theology.

We departed the base station at 9am the next morning with the temperature hovering around 2°F (-17 °C). Our plan was to follow the railway line (the old cog railway) to the top, a simple strategy to avoid getting lost. Every now and then we would sink into deep snow. About 11:30am we reached a small hut where we stopped for a short rest. The temperature was now -12°F (-24°C) and the wind chill around -50°F (-46°C). Shortly after reaching an elevation of 4,000′ (1,219m) Keith started to complain of frostbite symptoms in one of his hands, so we decided to call off the attempt and return to the base station. This was a mutually acceptable agreement as the cold was causing me considerable pain around the eyes and Mateus was suffering from leg cramps. I gave my giant foam outer gloves to Kevin who only had thin leather driving gloves. We descended the mountain at high speed, reaching the base station about midday. After lunching there we hiked back to the road. We relaxed at the Green Barn and accepted their offer of free accommodation for the night. Overnight the temperature dropped to -20°F (-29 °C).

The next day we departed around 9am and returned to New York via Boston.

1 Gloves
2 Sleeping bag
3
4 Billy
5 Foam or air mattress
6 Machete & sheath
7 Food
8 Maps
9 Nylon rope (2 types)
10 Crampons
11 Mess kit
12 Film
13 [redacted]
14 Raincoat or poncho
15 [redacted]
16 diary
17 Primus
18 Matches

chocolate
peanuts
dried fruits
bread
cheese
biscuits
rice
margarine
milo
milk powder?
salt?
sugar?
oatmeal
dried meat?

EURASIA
GREENLAND
T3
Barrow
NORTH AMERICA

The Arctic

1972

I was fortunate enough to spend five months on Fletcher's Ice Island (T-3) in the Arctic Ocean as part of Lamont-Doherty's geophysical program. Duties included maintenance of geophysical equipment, celestial navigation and daily weather reports.

18th May 1972
Fairbanks, Alaska

I departed New York at 11am on PAN AM 707 flight 801; flight lasted seven hours, cruising speed 966kph at 3,100' (9,500m). Weather was good and very little turbulence. Plane was half empty. We passed over Great Lakes, Niagara Falls, town of Uranium etc. Magnificent view of Canadian Rockies.

The plane landed in Fairbanks at 1:15pm Alaskan time. The weather was overcast and the temperature was 52°F (11°C). Airport is ten minutes ride out of the city. The population of the city is about 18,000. I took a room at the Hotel Fairbanks after a ride in from the airport in the local limousine service which serves as a bus and costs about $1.85. Driver takes you wherever you wish to go, and will pick you up in the morning for the run into the airport. Had a meal at Ken's Bar: two eggs, sausage, potatoes and coffee - $2.10. Did a little shopping – plenty of shops, particularly sporting goods. Quite a few Eskimos in town. Unsealed parking areas cause quite a bit of dust – town has general atmosphere of semi-frontier type.

Returned to the Hotel and wrote some postcards. Guys next door playing cards kept me awake for a while.

19th May 1972
Barrow, Alaska

Arose at 5am, packed gear then breakfasted at Ken's Bar. Temperature around 35°F (2°C) but it rose during the morning. Limousine driver arrived early but waited for me; there were no other passengers around at that time so we drove straight out to the airport.

Flight #1 to Barrow on Wien Consolidated Airlines 737 departed about 8:30am and lasted a whole hour and a quarter. Plane was pretty full – hunters, young couples, tourists, scientists (like myself) etc. We passed over frozen sea ice during the approach; it seemed to be about two to three miles (3-5km) wide. We walked out of the plane into the frosty Arctic air (about 7°F [-14°C]) and made a beeline for the airport lounge (in reality just a small shack) where half of Barrow had congregated. The group consisted of a large number of noisy Eskimo children and an equally large number of furry-faced SNARL (the term being my own corruption of NARL and meaning Scientists of the Naval Arctic Research Laboratory). I eventually came to the conclusion that someone was not going to meet me at the airport and so I got a lift to the NARL base with an Eskimo driving a truck. The ten-minute journey took us past the village of Barrow itself which is composed almost entirely of Eskimo families, and then onto the base a couple of miles further along the road. My first impression of the Eskimos was that they were friendly and pleasant.

The base consists of one large building complex and endless rows of semi-circular metal huts. The main building houses the living quarters, administration and laboratories. Although the base is situated right on the edge of the Arctic Ocean, the water is only visible during July and August; the rest of the time it is frozen solid (pack ice) and covered with a mantle of snow. Consequently, when we arrived at Barrow all we could see was an endless, flat expanse of white in all directions.

The main building also housed a very comfortable recreation room complete with billiard table, table-tennis table, radio and magazines. At least 200 men live in this building. The rooms were fully carpeted and well furnished.

One of the straps broke on my pack so I spent an hour sewing it back on. I was issued with some clothing – fur-lined parka, woollen trousers, thin cotton shirts, socks, gloves, rubberised boots and a sack to put them in. Took some photos of the base. In the hallways are photos of all phases of exploration and work in the Arctic. Meals were had in a cafeteria-style mess hall and the amount of food at one's disposal was enormous. I had a light meal and felt very self-righteous!

pan am flight 801
ticket no. 494959980

20th May 1972
T-3

Arose at 6:45am and went to breakfast where I dined on fried eggs, egg-bread, bacon, fried potatoes, orange juice and milk. The plane to T-3 was not scheduled to leave until 5pm so I spent the day reading Jules Verne. There are two hotels in the village which only open for the Summer with its small commitment of tourists. One of these is called 'The Top of the World'.

Our plane, a Lockheed Hercules, departed about 6:45pm with six passengers, countless fuel drums and propane cylinders. There were no seats initially so we just had to hang on as best we could during take-off; the cook and a metal ladder lost their equilibrium during this event, but no harm was done. After the plane had successfully made altitude we picked up the aluminium seats, placed them in convenient positions and rode out the rest of the flight in moderate comfort. The four-hour flight across the frozen ice to north of eastern Canada was uneventful but bloody cold! Even with my expedition parka on I was still shivering. No toilet facilities were available so one of the Eskimos peed on the floor. The rest of us just held on. A thick cloud cover obscured any view we might have had of the pack ice so we had no preview of our forthcoming exile. Touch-down was relatively smooth. Outside, the senses were blasted with deafening noise and fumes from the engine exhausts. The camp could be seen a few hundred yards away through a veil of wind-driven snow. All in all, a picture of complete desolation and lunacy! Shouldering my pack and clutching my other belongings I set off for the camp. Eventually I reached the mess hut, stumbled inside and dropped my gear.

The interior of the hut was somewhat ramshackle but warm and very inviting after my 'trans-arctic' trek. Seated around several large tables were a few of the inmates, drinking coffee and discussing the arrival. One of my co-workers, Isaac, was there, laughing at the new arrivals and in the process of getting himself 'well under the table'. He showed me our living quarters (Lamont's) and I nearly died from the stench and overall filth.

Our trailer was separated into a sleeping section and a lounge area. The only ventilation was the front door and all the windows had been blacked out to keep out the sunlight in the Summer months. There were two bunks against the back wall of the sleeping area, and another bed against one of the side walls. The beds were obviously never made and the sheets looked like the original unwashed articles (one guy had been sleeping in dirty sheets for over three months). Opposite one of the beds was a very dilapidated bookcase, covered with dust, old cigarettes, etc. The linoleum was peeling off the floor and loose tiles lay all over the place. The smell was probably the result of accumulated odours from hundreds of dirty socks, shirts, bodies etc. The lounge area was little better; the divan had been thoroughly chewed by Gunky (the resident husky dog) and its contents lay spilled all over the floor; old clothes, part of a weight-lifting set, more cigarettes and a host of other junk also lay scattered around. Isaac directed me to the closets where we stored our gear and told me to find one that was relatively empty. The first door I opened precipitated an avalanche of old boots, gloves, parkas and assorted clothes. In another I found part of a life raft, paddles, emergency rations and such-like. Eventually, by dint of considerable rearranging, I was able to store my gear away and retire for the night.

Jason, a co-worker from the United States Geological Survey (USGS) was scheduled to leave T-3 on the next plane so in the intervening ten days he instructed me in my duties. The Lamont crew usually work alternate 12-hour shifts and as Jason was doing the day shift, I worked the same, at least until Jason left. However, as the computer to the satellite navigator and the magnetometer were both down, much of that 12 hours was spent drinking coffee in the mess hut or at some other leisure activity. The three principal duties for the Summer were: a) celestial navigation which utilised the sun and a theodolite as well as the requirement for accuracy; b) recording weather data every three hours, and c) taking gravity observations from the gravity meter every three hours. In addition to these we had to change chart paper on all instruments requiring it (thermograph, barograph, precision depth profiler, anemometer and satellite printer), and keep the instruments operative, which meant repairing them if they broke down (luckily Isaac was fairly good at this). There were also numerous other small jobs which had to be done in the course of the Summer, like repairing the theodolite shack and insulating it – a little bit of carpentry which I was most proud of!

The camp consists of about 27 wooden and metal trailers, most of which are grouped fairly close together. The outlying huts are the five wooden hydro-huts, which house a metal A-frame and winch for obtaining samples from the ocean or ocean bottom through a hole in the ice directly beneath the centre of the A-frame. The huts include the Palace, which is the Lamont 'laboratory', the Colby Bay Country Club, which houses the University of Washington laboratory, and the shack out by the runway bearing the sign: 'T-3 International Airport'. The central area of the camp includes the mess hut, the seven sleeping trailers, the generator shacks, weasel shack (a weasel is a sort of car on tracks), the various 'shitters' – both one and two-holers, several storage shacks, and lastly, the old weather hut which has a large fibreglass dome on top.

The camp is maintained by the Naval Arctic Research laboratory at Barrow which provides a station leader, several Eskimos to do the mechanical and maintenance jobs, a cook, and on this occasion an assistant to help the cook – washing dishes etc (the cook's son in this case). The rest of the personnel are scientists representing a small group of universities from mainland US. Of these, Lamont and the Universities of Alaska, Idaho and Washington have permanent quarters here. Personnel from these institutions stay anywhere from ten days to over a year on the island, though the normal period is from three to five months. During Winter, which extends from 1st September through to 1st May, planes land every ten days; in Summer, planes are unable to land on the soft snow and small lakes which characterise the runway at this time, and consequently supplies and incoming mail are dropped by parachute. Paradrops usually come every two weeks though bad weather and 'other commitments' often delay them. Thus personnel who stay for the Summer have to remain for at least four months. Only in Winter can a person make a very short stay.

During the ten days before the final plane departed I familiarised myself with the camp and the surrounding landscape. The depth of snow cover was about 3' (1m), though deep drifts were piled up everywhere. The snow was quite firm to walk on, except that which had freshly fallen. The camp is situated near one edge of the island so it is an easy matter to walk out onto the pack ice. Away from camp the atmosphere is entirely different; the stink of diesel fuel is gone and the noise from the generator is greatly reduced.

Peacefulness and quiet beauty are united as one and transcend all other impressions. Time stands still out there on the pack ice, everything is real and all that is not concerned with life and death is irrelevant.

You won't find any frilly poodles out there. The pack ice is about 10' (3m) thick as compared with about 100' (31m) for the island. Along much of the contact between pack ice and island, pressure ridges develop. Around T-3 they are typically small and can be climbed or crossed easily. The highest point around T-3 is a 25 foot ice pinnacle (8m) about a mile from the camp which we named 'The Pinnacle'. A few days after arriving at the base one of the temporary residents suggested we go out and climb it. So my co-worker, Stew, and I collected an ice-axe each and walked out there, crossing some old pressure ridges on the way. The snow was already starting to get soft and it was a good half an hour before we reached our objective. One side of this block was almost sheer so we decided to leave that face until we knew what we were doing with our ice-axes. Around the other side the going was easier though we still had to cut steps in the ice. From the top we had a spectacular view of the surrounding pressure ridge. Behind the pressure ridge the pack ice displayed a jumbled array of small ice mounds and peaks, something like a chessboard in mid-game receiving a mantle of snow, and then freezing. We descended the way we had come and then headed back to camp.

29th May 1972

On the night before the plane was due to arrive, Stew and I returned to the Pinnacle to make a climb on the steep face and to take some photos. With the camera mounted on the tripod and being manned by Stew, I tackled an overhang along one edge of the face, first cutting steps diagonally across it and then securing the point of the ice-axe near the top and hauling myself up. Fortunately the axe held and I wasn't pitched into space as my dedicated photographer was hoping. After taking some more photos, we returned to the camp.

30th May 1972

The next evening the plane arrived, bringing six guys from the University of Washington (UW) (five of whom were volunteers), two Eskimos, one guy from USGS and the cook's son. This brought the number of permanent Summer residents up to 15, including: station leader (Ted); Eskimos (Ron, Fred and Joe); co-workers (Isaac and Gavin); the UW contingent ("False Mike", "Weird Mike". "Gronk" [Frank], James, Daniel and Karl); from USGS, Jason; and finally the cook and his son.

After the plane had departed and everyone had either gone to bed or occupied themselves in some way, Isaac and I quietly left camp and set off for the old abandoned Gravity and Magnetic (GM) camp, about a mile away at the other end of the runway. The camp had been declared off-limits as there was supposed to be classified material there, but the prospect of rummaging through an abandoned camp that apparently hadn't been visited in over two years was too tempting to dismiss for purely bureaucratic reasons. So, it came to pass that around 2am, after a journey of approximately three-quarters of an hour, we found ourselves amidst a small group of wooden buildings, all of which were partially buried by huge snowdrifts. There were about eight buildings in all, of different shapes and sizes, so we decided to try the largest one first and then work our way around to the others in turn. Although we had an ice-axe and a small crow bar with us, most of the doors did not require such drastic measures. Several of the doors had locks on them but none of these locks was actually closed.

The first hut proved a veritable treasure chest as we unearthed from the snow-covered rubbish several pairs of snow-shoes in excellent condition, two sets of ski poles, weight-lifting equipment, water-proof hip boots and a somewhat battered sled. We claimed a pair of snow-shoes and ski poles each and left the rest of the stuff for another time. The other huts were not as lucrative as the first but contained several items of general interest such as books, which we planned to come back for. We entered the mess hut through a window over the door, the latter being completely drifted-up and frozen-in. Inside we found a table laid with food, plates and cutlery, a refrigerator stocked with decomposing food, shelves laden with boxes of cereal, a medicine cabinet full of old bottles of medicine, bandages, magazines and numerous books etc. A layer of ice covered the floor and snow drifts had piled up in many places. The snow had originally blown in through cracks around the door and windows before melting in the Summer and then freezing in the Winter to form the ice. Not all the huts had been subjected to this kind of invasion however; the sleeping quarters was completely free of any ice or snow and the place looked just like it must have before the GM personnel moved out. The bedding was still on the mattresses and the bunks were in excellent condition. Apart from pictures and pin-ups on the walls, little else was to be found there. The main work hut or laboratory was also free from snow and ice but was a shambles nevertheless. Radio and electronic equipment were lying around with wires going in all directions. Weather recording instruments were also found there. Having completed our reconnaissance of the GM camp, we strapped on our newly appropriated snow-shoes, clutched our ski poles and set off back to base.

7th June 1972

On the morning of 7th June, I made my third visit to the Pinnacle. This time four guys from UW came along and we made a fairly extensive tour of the larger ice peaks in the vicinity of the Pinnacle. The weather was lousy so we didn't try any climbing.

On 8th June, and again on 10th June, I trudged back out to the GM camp, in the company of several guys from UW. We found the sled and filled it with a pile of goods from the abandoned huts. We brought the weights back for our proposed weight-lifting room in the old weather hut at our camp.

About this time, Isaac and I decided to do something about our disgusting living quarters, so for the next few days we were busy cleaning out the junk and painting the walls and cupboards. The only paint in camp was yellow, so by the time we had finished, our trailer resembled something like a hospital ward. I threw out my old mattress and frame as the bed had a decided lean to it, and replaced them with three new mattresses which I piled on top of each other on the floor. We ripped up all the old tiles off the floor in preparation for painting the floor at a later date. The unsightly 'stuff' covering the windows was removed and rough curtains put up in its place. The vents were opened up and fresh air allowed to circulate for the first time in many months. The divan we covered with a blanket to hide the holes. The first phase of the clean-up was completed in three days. We still had to paint the floor, and the walls of the front room, but even without this, the transformation was quite amazing.

The old weather shack was also undergoing remodification at this time. We had decided to convert it into a recreation hut, containing a library, weight room, and table-tennis table. The books for the library we scrounged from all the trailers around the camp, as well as from the small library at the GM camp. The weights came from our trailer and the GM camp (later on we found some more weights in another trailer in camp). The table-tennis table was eventually built around the end of July, and the net and bats etc. we obtained from Barrow via a paradrop. We put all the weight equipment in the upper room of the weather shack, which was enclosed by the dome. The library and table-tennis table occupied the downstairs room.

15th June 1972

Another landmark near the camp which I had a strong desire to visit and photograph was the old, wrecked USAF C-47, which lay on the other side of the runway, almost completely drifted over. The only exposed parts now are the nose and one of the wings. Old photographs taken by Phill (a co-worker of my Antarctic trip) in 1963 show the complete plane, with very little snow cover. Even during the warmer parts of Summer the amount of exposed surface hardly increased. In a few years the plane will probably be permanently covered.

Towards the end of the month when the temperature started to go above freezing, small meltwater streams materialised around the camp. These streams fed a rapidly growing lake right in the centre of camp, as well as several smaller ones around the camp. The NARL crew dug an outlet canal so the large lake would drain off behind the camp where a drain hole penetrated the sea ice, thus permitting the lake water to flow back into the ocean. As the Summer progressed, most of the snow around camp melted and ended up back in the ocean via this and other drain holes. Further away from camp, the meltwater formed large ponds which eventually froze over when the temperatures dropped below freezing. A considerable amount of ice also melts during the Summer months but this loss is probably compensated by ice accretion at the bottom of the island.

All the lakes around camp contained a considerable amount of diesel fuel, grease, oil etc. which did nothing to improve the already squalid appearance of the camp. As the snow gradually melted down, piles of old garbage and the burnt out remains of former buildings became exposed. A common policy in camp is to throw out anything that doesn't work or isn't needed; up here that means literally throwing it out into the snow and who cares if the place looks like a garbage dump in Summer – in Winter everything around the camp is covered with snow and conveniently out of sight. The T-3 garbage dump itself is a spectacle which has to be seen to be believed. It stretches endlessly; a sea of rusty cans, cardboard cartons, food scraps, old boots, burnt timbers etc., undecomposed and waiting for the day when warmer waters will melt the island and deposit it all in the ocean. Also waiting are several thousand 55-gallon drums, relicts of earlier days, before pollution control became a byword. One side-effect of contaminants in melt ponds is the lowering of the freezing point of the water, so that while lakes on the outskirts of the camp were freezing over at the end of Summer, most of the lakes in and around camp were still unfrozen. This became quite a hazard after a good snowfall and several unwary persons, myself included, found themselves struggling to stay vertical in a chaotic mess of broken ice, snow and oily water, which usually came up to the waist, if you didn't fall over. I did, but was supported by the ice so the top part of me stayed dry. On this occasion the water had frozen but the NARL crew had broken it up by driving the Cat across it.

19th June 1972

The first paradrop took place about 6pm, out on the runway. Just before the plane was due to arrive we all rode out to the runway in the two weasels and gathered at a point just to the side of it. From this vantage point we watched each drop, without fear of seeing a free-fall directly overhead. When the plane arrived, it made several preliminary passes before actually discharging its cargo. Most drops occur from a height of around 300 feet (90m) so that the pilot can see where everybody is, and so avoid a calamity. After the preliminary runs the pilot then proceeded to make three drops at intervals of about five minutes, thus giving the retrievers enough time to collect the boxes after each drop. A strong wind was blowing and as soon as a crate hit the ground it would take off at a fair pace behind a billowing parachute. The trick was to grab the centre of the chute from the outside, swing it around into the wind and twirl it around itself so that it couldn't reopen. We then removed the chute from the box and hauled the box into a large steel sled. After the first drop, everyone became disorganised and people were spread everywhere. Consequently the second load nearly claimed some casualties – a large crate came swinging down on a collision course with one of the weasels which was immediately abandoned by all on board (me included). The crate missed the vehicle by no more than a foot and thudded into the snow a few feet further on. Needless to say we regrouped for the last drop and watched the remaining nine boxes plunge to earth. The crates took only about ten seconds to cover the 300' (90m) distance to the ground. No freefalls occurred out of 27 parachutes which was rather unusual according to some of the old hands.

When all the crates had been loaded onto the sleds they were towed back to camp for disembowelling. Everyone set to work with crowbars and wire-cutters and in less than half an hour all the contents had been disgorged and the mail sorted and distributed. I had a field day where mail was concerned – 17 letters. I felt sorry for those poor souls who received only one or two – receiving mail up there on the ice means far more than a lot of people imagine, if they stop to think about it. Mail is the only real link with one's own personal reality and serves to reduce the isolation to a more tolerable level.

Four films came in the drop so later on that night we showed two of them: 'What's New Pussycat' and 'Where Bullets Fly'. The Eskimos had received several bottles of whiskey so they proceeded to get themselves thoroughly drunk during the night. When I entered the mess hut the next morning, one of them was stretched out on the floor and another was gazing vacantly into space. This was to become a familiar pattern during the following months. Immediately after every paradrop they would get themselves plastered and then sleep it off the next day. Relations between the Eskimos and the other personnel were extremely good between paradrops so there was rarely any trouble when the alcohol did flow. On one occasion, one of the Eskimos drove the Cat through a wall of the station leader's hut. It was apparently an accident - albeit one caused by drunkenness - and was no big issue. However the peaceful conditions prevailing this Summer appear to be the exception rather than the rule. The only sure way to remove this is to make T-3 dry.

27th June 1972

I went out to the GM camp alone to bring back a bag full of frozen gloves, in the hope that I might be able to salvage a respectable looking pair from the assortment. When all the gloves were dry I found not one, but several pairs in quite good condition. These of course I kept.

A brief period of good weather the next day prompted a few of us to make another visit to the Pinnacle. This was primarily for the purpose of taking some photographs, but also for the exercise. Frank and I wore snow-shoes while James came along on skis. After a brief sojourn at the Pinnacle, we continued on for about half a mile further through a sparse forest of small ice-mounds.

When conditions are overcast, all contrast disappears from the surface of the ground and the snow and sky merge into a pale grey homogeneity. Under these conditions, the Arctic is dull, drab and decidedly boring. With the appearance of the sun however, the vision is transformed into an intricate blending of light and shadow. White becomes pure white and textures take on an intense complexity. Light rays dance and sparkle off snow and ice crystals. The Arctic becomes something alive and beautiful.

Around this time, many of the trailers in camp were becoming pedestalled as the surrounding snow and ice melted away. When an alternative site had been prepared, the building would be towed off its ice pedestal and placed on this new site, and the pedestal graded down to ground level. This process had to be repeated several times in the case of the wash-hut, but no more than twice for most of the trailers. Ours was only removed once, and consequently the building was a good two feet (60cm) above the surrounding ice surface by the time the de-pedestalling took place. The trailer had to be dropped into a shallow lake behind the pedestal so the latter could be graded away. The trailer was then towed back into position.

4th July 1972

Went para-sledding with several UW guys out on the other side of the runway. As the name implies, this sport involves a sled being towed behind a parachute, or two parachutes depending on the size of the sled and the number of people riding on it. The sleds never move at more than a few miles per hour, but being so close to the ground, the impression is of a much greater speed. The art of starting a para-sled run is worth mentioning. First the parachute is laid out behind the sled so that the cords can be straightened without the wind hindering the job. When the chute has been arranged satisfactorily, one of the riders will take the top of the parachute and swing it around so that the wind can billow the chute open. Then as the chute opens and starts to pull the sled, the chute-opener has to chase the moving sled and jump aboard. When I tried this stunt I found the only way to get on board was to dive at the sled and grab the retreating edge of it with my fingertips. After being dragged face down in the snow for several yards I managed to pull myself up and roll onto the sled. This feat is also necessary if by chance one of the occupants falls out of the sled during the run. We ended up having a race, James and I with one parachute towing, against three other guys in a sled using two parachutes. As we drew abreast we hurled snowballs at each other, followed by a more vigorous personal assault. The race ended very abruptly when the sleds piled up on a mountain of oil drums at one end of the runway.

We rode back to camp in one of the weasels, towing behind us one of the sleds with two occupants. I was one of the occupants on the trip out and the sensation is one of hurtling down a bobsled run behind an equally fast snow plough whose spray nozzles have been turned to the rear. As if that wasn't bad enough, the guys in the weasel up ahead kept throwing things at us, so we retaliated by scooping up snow and lobbing it into the vehicle. That night we celebrated 4th July with an outdoor barbecue, staying outdoors just long enough to cook our steaks!

The paradrop scheduled for 5th July came on the 6th, owing to bad weather here the day before. Fourteen boxes were dropped and all the chutes opened properly. I received six letters. A dartboard came in the drop so we promptly put it up in the Pleasure Dome and availed ourselves of it.

11th July 1972

On the evening of the 11th, tempted by the good weather which had been lacking all month, four of us decided to take a long walk out to one of the rock piles on the far side of the island. At several localities around the island, small deposits of moraine material, typically strongly frost-shattered, became uncovered during the Summer melt. The rock pile we were heading for was about 5.5 miles (9km) from camp near a pressure ridge marking the edge of the island in that direction. Daniel and False Mike were not wearing snow-shoes so they were soon following a more round-about route over firmer ground, about 200 yards (180m) abreast of us. Frank and I spotted a small rock pile near the main one so we headed there and examined the material. We could see Daniel and Mike quite a long way off heading for the large rock pile, like two ants crawling across the surface of a cream cake. We decided it would be too much like hard work to visit both rock piles so we spent an enjoyable half hour looking at the rocks and basking in the sun. The rocks all seemed to have been derived from the same source, probably the result of frost-shattering breaking up an originally homogeneous boulder. The others reported the large rock pile to be very inhomogeneous, obviously reflecting a multiple source origin for those rocks.

On the way back we stopped to investigate an abandoned hut about a quarter mile from our rock pile. When we reached the hut we found the door frozen in and the windows too narrow to squeeze through. After unsuccessfully trying to smash the door down with our shoulders we resorted to a technique whereby we grabbed hold of the door-frame with our hands and pushed down on the top of the door with our feet; a resounding, splintering crash informed us that our technique had succeeded, with the top part of the door lying amid the rubble on the floor inside. We entered the hut and examined the traces of former habitation. A layer of ice covered the floor and snow was everywhere; everything that was in any way connected with the floor was firmly frozen in. A rusty old Coleman stove lay on a table in one corner of the hut. Electrical wires were draped across the ceiling and miscellaneous electronic equipment was scattered around. A large wooden box on the floor held a variety of things and included a partially frozen sleeping bag. Shelves along the walls held a variety of foodstuffs, most of which were in packets and appeared to be unfit for human consumption. There was virtually nothing worth taking so we left after a short while and headed back to camp. The trip back took considerably less time than our outward journey, probably a function of our greater adroitness at using snow-shoes. Every now and then one of us would catch the top of a snow-shoe in a small snowdrift and precipitate himself forward onto his knees. In this position it is no easy matter to right oneself and quite a bit of time is wasted in manoeuvring one's limbs into the normal positions.

On all our treks, Gunky used to follow us faithfully, usually at some distance ahead. As the Summer wore on, and the snow became softer, the poor animal found the surface offering her paws very little resistance and she would be continually falling through. The poor beast would struggle around on its stomach until it could get some leverage on firmer ground. After a while she got smart and started placing her paws in our steps. It was quite a joke to see Gunky stepping it out right behind someone's heels.

Over the next few days, Isaac and I painted the rest of our trailer. The floor throughout is chocolate brown, which matches well with the light blue walls of the front room and the yellow walls of the sleeping section. The smart interior of our trailer contrasts markedly with the outside appearance. The aluminium surface is splattered with paint and adorned with numerous clichés and signs ("Have a nice day", "Fascist pigs", peace signs, etc).

The most colourful building in camp is a small wooden hut painted with red and black squares, which houses the "shitter". From about the end of June to early August this was in continuous operation. (Before that period we used a small tin-can-with-a-seat affair in one of the unused trailers; after that period we used a modern-looking, two-seat arrangement utilising two 55-gallon drums, built by the station leader). As the ice and snow around this red and black hut slowly melted down, its hold on the ice pedestal beneath it became precarious at best. Whenever a strong wind was blowing, the hut would rock back and forth, daring courageous (or desperate) souls to try their luck. No-one who has ever sat atop a 55-gallon drum, three-quarters full of sh.., with the whole structure swaying menacingly from side to side, would ever again complain about lavatory facilities. Such insecurity maketh cowards of the bravest of men! That shitter did in fact claim one victim; Daniel was apparently perched up there when the whole caboozle flipped over onto its side. He was lucky, however, and managed to avoid a fate worse than death.

17th July 1972

Slight drizzle coming down but the wind was good so we decided to get in some more para-sledding before all the snow melted away. The walk out to the runway and the start of the run was a slow, wet and exceedingly laborious process. At each step we would sink down to the top of our hip boots, sometimes further. When the snow and slush became too deep for comfort, we would crawl forward on our hands and knees hoping that the increased surface area in contact with the snow would support us more adequately. This usually resulted in wet gloves and hands. Frank and I were not anxious to spend hours walking in such a fashion, so when we had gone a reasonable distance we decided to stop there and set up the parachute for the run. The other three guys were all gung-ho so they continued on for quite a way. The cords on our parachute were slightly tangled but we managed to get it up and pulling. The others were nowhere in sight so we sailed back to camp with only the swish of the sled gliding over the snow for company. Our ride came to a halt at the pile of drums that had stopped us previously, and we waited around for the others to come in, which was a few minutes later. The second sled missed the drums and continued for a short distance past them, coming to rest at the base of a small snow slope.

The next day heralded the worst weather of the Summer for T-3. A fierce blizzard lashed the camp for most of the day, with wind speeds up to 51mph (82kph). The temperature hovered around 30°F (-1°C). Although the sky was overcast for most of the Summer, storms were rare and snowfalls generally light. With 24 hours of sunlight during the Summer months, the average number of hours of sunshine per day during July and August did not exceed 2.5. This contrasts with over 12 for April and the early part of May. June's average was about 3.5. The highest temperature recorded was 38°F (3°C) in June. From the latter part of June to the first week in August the temperature rarely dropped below 30°F (-1°C). Towards the end of August the temperature started to drop but even in the early part of September the temperature rose above 30°F for brief periods. On 10th September the temperature hit 0°F (-18°C).

On 19th July a twin-otter with nine members of the Royal Mounties on board landed on T-3, on wheels! They stayed two or three hours then took off again after the runway had had some minor repairs done to it. Apparently the plane had dug into the snow pretty deeply and then got stuck at one end. I was asleep when the plane landed (I work night shift) and remained asleep even though Isaac claimed to have woken me up. I must have answered him in my sleep as I had no memory of the incident at all. So I missed everything, including the badges that were handed to most guys here. To make matters worse, the same day the USAF informed us that the paradrop scheduled for the 21st was cancelled and that we would have to wait another two weeks for the next drop. I was p…off!

Lousy weather continued until 27th July when the sky cleared for a few hours. Present position: 84.10.2 N, 83.56 W. In the short term, the island moves according to the prevailing wind direction; over a long period the dominant influence is attributed to ocean currents. The present trend of these currents is south-east, but as they are cyclical currents the eastward component will reverse direction shortly. A fix on 20th September gave the following co-ordinates: 83.45.2 N, 88.51 W. This position is approximately 70 miles (110km) off the north-east coast of Ellesmere Is. The speed at which the island moves is a function of the wind speed, current speed and the cohesiveness of the pack ice around the island. Most of the island is permanently attached to pack ice though leads can develop in the pack ice further out. The leads tend to be long, narrow breaks in the pack ice though some over a hundred yards wide are occasionally seen. Typically they open and close fairly rapidly, so beware anyone who gets caught on the wrong side of one.

27th July 1972

This was the day of the great naval battle on T-3. By this time there were several large and fairly deep lakes around the camp and it was suggested by someone that an interesting diversion might be a life-raft race on one of these lakes. The very pronounced, but friendly, rivalry which existed between Lamont and the UW, prompted these two groups to take up the idea, and preparations were soon under way. To even up the numbers, Lamont recruited one of the NARL guys and the cook's son to their cause. Even so, the UW still had an advantage of two. It soon became clear that the race was to become something more than just a race, with both groups threatening to attack each other when the two met on the 'high seas'. During the few days before the race, repairs were made to our raft and both sides secretly stockpiled ammunition.

Finally, with the advent of some good weather on the 27th, the entire camp personnel walked over to a sizeable lake behind the Palace for the start of the historic battle. The six members of the UW team launched their brand new, ten-man raft and promptly disappeared inside the canopy which covered their raft. The four-man Lamont team launched their raft almost simultaneously and threw flour bombs at the retreating backs of their opponents. The floor of the Lamont raft had exploded (too much pressure) just prior to the launching, so we found it exceedingly difficult to pole the raft while standing on its limp, uninflated lower floor. Nonetheless we were able to move about the lake with considerable dexterity, shouting taunts at our opponents who were struggling to manoeuvre with wooden and aluminium paddles. The UW crew all wore canvas hoods over their heads and brandished wooden swords and shields, to the accompaniment of fierce war cries and blood-curdling yells. Somewhat taken aback by this display of barbarism and feeling more than a little defenceless having completely extinguished our supply of ammunition, we turned tail and sought to put some distance between the two vessels. Luckily their locomotion was not rapid and we were able to keep a fair distance between us.

At this point, our adversaries opened fire with a surgical tubing mortar, sending an assortment of vegetable products flying in our direction. By a stroke of fate, their first projectile, a raw egg, found its mark squarely on the back of my head. I was wearing a cap at the time and was leaning out over the ledge of the raft so that the force of the impact lifted the cap off my head and sent it flying into the water. Recovering from the shock, I turned around and was greeted by a loud cheering from the other raft. Spurred on by this initial success the UW crew fired several more volleys but luckily all fell wide of their mark. When all ammunition had been spent, the two rafts drew close for a hand to hand encounter. The UW guys grabbed the ends of our poles and pulled the two rafts together amidst wild cries on both sides. As soon as the rafts touched, our opponents leapt into our raft and a great struggle ensued for several minutes. Although outnumbered, we held our own and managed to throw one attacker into the lake, without suffering any casualties ourselves. After about three minutes a truce was called and the rafts were propelled onto the shore. The guy who had the honour of being the only one to be immersed in the lake was again dunked for stating that he could not be any wetter. After pulling the rafts up onto the shore, everyone retired to the mess-hut to recuperate.

3rd August 1972

Paradrop finally came about 9am but the first load accidentally fell out onto the pack ice about 7 miles (11km) from camp. The second load landed safely within reach. During the next couple of weeks, several attempts to locate the first load were unsuccessful. Six packets of mail were lost in those boxes.

Average daily sunshine for July was 2.5 hours. This compares with 3.4 hours for June. Average temperature for July was 31.5°F (-0.3°C). About this time, Isaac and I painted the inside of the Palace – orange, blue, purple and black! We also built some cupboards and generally straightened up the place.

During July and August, the surface of the 'ground' was largely devoid of any snow cover, and much of the island ice melted down to produce the texture called candle ice. This type of ice is characterised by jagged, finger-like projections caused by irregular melting patterns. Gunky found this type of ice very painful to walk on and for several weeks she nursed bleeding paws.

On 6th August, False Mike and I walked over to a rock pile behind the dump, about a mile from camp. The area was studded with small lakes and the surface of the ice was buried under wet mushy snow, so by the time we got back, our legs were soaked and very cold. Most of the rocks were volcanics and gneissic metamorphics. The trip to and from the rock pile brought us into close contact with the dump which provided us with a drier passage for a good part of the way. The very cold temperatures experienced in these latitudes prevent decomposition, so metal cans and food scraps lie side by side, oblivious to the ravages of time that accompany warmer climates.

1.61 km = 1 mile

8th August 1972

About 3am the cloud cover which had been present for several days finally cleared, exposing a perfect sky and a promise of at least several hours' clear weather. After taking a sun fix, I went over to the UW lab and managed to persuade Daniel and Mike to make preparations for a recovery expedition to the pressure ridge on the opposite side of the island where we suspected the lost paradrop boxes might be. After some hasty preparations, we set off – Daniel, Frank, Mike and myself. We carried a rope, tools for opening the boxes, a rifle, walkie-talkie, ice-axe, some emergency equipment and a couple of cans of drink.

Right from the start, we found ourselves wading through or crawling across snow-covered melt-water lakes. The snow cover was usually thick enough to support a body in the prone position, so whenever the slush came up to the tops of our hip boots, we would look for some firm snow and cross on our bellies. As a result of numerous slip-ups, my trousers and socks became soaked not long after the start. Between lakes the surface was generally hard and largely bare of snow cover, which provided for easy walking.

Three hours of painful and arduous hiking brought us to the pressure ridge on the opposite side of the island. This ridge was very narrow and quite low and consisted of upthrusted blocks of ice with considerable drift accumulations on the sides. Small blue melt lakes dotted the flat land at the base of the ridge. As there was still no sign of the boxes, we split up into two parties and selected appropriate vantage points for studying the pack ice on the other side of the ridge.

Just beyond the ridge, a long narrow lead paralleled the ridge for as far as the eye could see. The lead was about 15' (4.5m) across and in places was choked with an assortment of angular ice blocks, one such block forming a convenient bridge. There was no wind and the ice on the other side of the water looked stable so we decided to risk crossing the bridge and continuing our search on the other side. We roped up and Daniel walked cautiously across the ice bridge, belayed by myself up on the ridge. When he had safely reached the other side, I followed across. Our progress across the lead was watched by our two companions who were standing some distance away on a high point of the ridge. The going was considerably easier on the other side of the lead and we were soon among the jagged ice hummocks which typically dominate the surface of pack ice. We climbed the highest ice hummock we could see and scanned the horizon thoroughly for a trace of the boxes, but to no avail. There was nothing but ice and snow stretching to eternity in every direction. The only sound to be heard was a soft crunching sound, apparently a long way off, and attributed to either ice under pressure or a polar bear walking across the surface.

With an open lead behind us and the possibility of a polar bear in front of us, we decided that any further progress forward could be detrimental to our health and so we turned around and began the long journey back.

We rejoined Mike and Frank on the other side of the ridge, but waited for Gunky to come over to us before starting our journey back. Her feet were cut from walking over candle ice and her progress was extremely slow. When she finally arrived, the others wanted to make 'booties' for her feet before heading back, but I objected on the grounds that the booties would not stay on long enough to warrant the time and effort and that, however slow her progress, dogs were a lot tougher than humans and she would make it back okay.

My real reason for objecting, which I didn't voice, was that my feet were freezing and *had* been for most of the three-hour journey to the ridge. My left foot wasn't bad but my right foot was very much wetter and consequently refused to warm up. Five minutes after we started our return Gunky's shoes started to come off. When all the shoes had come off we waited another 15 minutes while they were put on again. Aside from the delays associated with Gunky, we made much more rapid progress back than on our outward journey, largely due to the experience we had gained in crossing the numerous lakes which blocked our path. Even so, I still managed to get more water in my boots and by the time we reached our last lake my right foot was completely numb. I was starting to complain about the slow progress the others were making (waiting for Gunky etc.) and received much unneeded heckling in return. Only one of the others had obtained wet feet and he had sustained that condition just a few minutes earlier. When it had happened he said to me: "how long does it take for feet to warm up again?", and he was really taken back when I replied: "they don't". He then asked me why I hadn't been complaining about my feet, and I told him that I assumed everyone else was suffering with the same problem. When we reached the last major lake and the others stopped to wait for Gunky, I decided that I needed to finish the trip on my own, at my own pace.

I was soon a long way ahead of them and eventually entered the camp a good 20 minutes in front. The first thing I did upon returning was go to the wash-hut and place my foot in a basin of hot water. This produced an extreme itch in that foot accompanied by swelling and the development of black and purple blotches behind the toes on the under-surface. This discolouring disappeared after about ten minutes but the swelling remained for two or three days.

14th August 1972

Weird Mike and I walked out to the old Airforce camp, a trip of about three-quarters of an hour each way. The remains of the camp looked just like a big rubbish dump with only one recognisable building. Even that was pretty far gone. Our main purpose in going to the camp was to bring back some helium balloons but in addition to these we also found a set of crampons which needed considerable repairing but were nevertheless a valuable find. We also brought back a rope-and-tackle affair which Mike had found. We were able to walk directly on ice much of the time as cooler air temperatures had caused many of the unpolluted lakes away from camp to freeze over.

The next day, Weird Mike and I set about inflating the balloons we had brought back. The balloons first had to be immersed in hot water which makes the balloon supple so that it will stretch evenly when inflated. The first balloon burst when it bumped against a sharp projection on a helium cylinder, but our second attempt worked and we ended up with a large balloon, about 7 feet tall (2m). The balloons we normally used for establishing ceiling heights rarely exceeded a couple of feet (60cm) in length and had virtually no lifting power. We found that our large balloon could lift about five pounds quite comfortably. We tied an empty Pepsi can to the neck and released the balloon. It shot up immediately, flattening quite considerably due to the air pressure on the top of it. We watched it for several minutes before it finally vanished through the clouds, a tiny yellow speck on its celestial journey.

Another paradrop arrived on the 15th and was very successful though there wasn't much mail. However one box was full of candy bars, life-savers, other assorted sweets and several jigsaw puzzles, all of which were generally beneficial to our morale (particularly mine).

The sun was now 6 degrees off the horizon at its low point but at least 18 degrees at its highest point. Starting to get some sunsets now. Temperature alternates around 20-30°F (-7 to -1°C), depending on whether the sky is overcast or clear, respectively. Weather synopsis for August: average temp 29.8°F (-1.2°C); max temp 36.2°F (2.3°C); min temp 16.4°F (-8.7°C); number of hours sunshine per day 2.65; max wind speed 19 knots (35kph).

31st August 1972

Paradrop arrived on the evening of the 30th. I received ten letters and a jigsaw (title: The Sensuous Woman). The puzzle was a big hit with all the guys – it was a nice change in the monotony of life up here. Have had a mild toothache since 28th August but am hoping for the best. Still weightlifting with Daniel at 6am each morning. Bench press 160lbs (~73kgs); Military press 130lbs (59kgs); curls 80 etc. Daniel is benching around 200lbs (91kgs). Sun first dipped below horizon on 10th September, but only stayed down for a few minutes. Temps getting colder now – 10 to 20°F range (-12 to -7°C).

12th September 1972

Last paradrop came today. Received second jigsaw. That night we put on one of the films which had arrived in the drop and everyone sat around drinking whisky or T-3 shandies (ethanol and 7-up).

During the last couple of weeks on the island our celestial fixes indicated a fairly rapid island movement of up to seven miles (11km) a day in a southwest direction. This indicated loose pack ice somewhere ahead of the island. Although the temperature was dropping steadily, the sky remained overcast most of the time, dispelling the theory that colder weather brings clear skies. This may be the case elsewhere or at different times, but this Summer the clouds stayed right there.

The toothache vanished about the middle of September to my great relief.

28th September 1972

The first plane finally arrived about 9pm for our departure. It was dark out at this time and the sight of the plane coming in between the landing lights of the runway was a memory to be preserved. When the plane had been unloaded and reloaded, and the crew fed, the departing passengers said their farewells and clambered aboard.

I had to carry the gravity meter back to Barrow, so I made it fast and then found myself a seat for the four-hour flight to the town. The plane was a four-engine Hercules that was not meant to carry passengers, so conditions were not very comfortable. There were no seat belts and the seats were not even bolted to the floor. The air pressure inside the plane was equivalent to an altitude of 15,000′ (4,500m) and the heaters were turned up far too high. It goes without saying that we were not offered any food during the flight. Everyone cheered when the plane finally took off; and cheered again when we landed at Barrow.

Postscript: The island didn't last much longer. The station remained active until October 1974 and was last visited in 1979. It was last seen in 1982, easily identified by the remains of the C-47 plane that had crashed early in the island's history. T-3 is believed to have moved into the Atlantic Ocean via the Fram Strait sometime in 1983 where in the warmer waters it eventually melted.

1 knot = 1.85 kph

Above: Barrow, Alaska.   Below: T-3 camp.

Above: Para-sledding   Below: Isolation.

Above: Wrecked USAF C-47 plane. Below: The abandoned hut.

MEXICO
Volcan Popocatepetl

Adventure to Volcan Popocatepetl

November 1972
Mexico

When we arrived in Mexico City I enlisted the aid of an English-speaking passenger to help me find the right bus to Popocatapetl volcano (El Popo). I went to bus line Cristobal Colon and booked a passage to Popo Park, a small village near the base of the mountain. San Pedro might have been better but I wasn't sure any buses went there. The trip took about two hours and cost 7 pesos (55¢). At Popo Park there is a sort of hotel which rents cabins for 80 pesos a night. I talked them down to 50 pesos and even managed to have a cold shower. Had a meal there too.

Early on the morning of Monday 13th, I set off along back roads and shortcuts towards the mountain. The other mountain, Ixtacuatl (The Sleeping Lady) is very close and a little lower than El Popo. On the way I managed to pick up some food and filled my water bottle with mineral water. Most of the day was spent stumbling through cornfields looking for decent paths. I eventually came to San Pedro where I consumed two large bottles of soft drink in smart fashion. Some of the farmers tried begging but without success! I had trouble understanding their Spanish, particularly when they said go right and pointed to the left. Well, I followed their arms and eventually came to a sealed road that was heading up the mountain. So I plonked my pack down on the side of the road and waited for a lift.

A little while later a group of mountaineers from Orizaba came by and gave me a lift. There were 11 of them altogether, in two cars. The road continued for quite a long way before terminating at a parking area and bungalow at around 14,000′ (4,500m). They all had crampons and ice-axes and very rough, makeshift protective clothing. One guy's raincoat consisted of a plastic sheet with a hole for the head and was tied around the middle with a piece of string. They were very friendly and invited me to climb with them. We entered the bungalow around 6pm and had a light meal in front of an open fire. Shortly after we arrived, an American guy arrived by himself and joined our group. He and I shared a room in the bungalow that night, at a cost of 6 pesos each. He lacked the necessary hardware to climb with us so agreed to stay and look after my camera and pack. He would then borrow my gear when I had finished with it and attempt the climb himself.

We departed for the summit at 2am the next morning in order to climb while the surface of the snow was firm. I was warmly dressed with a down parka, flannel shirt, mesh undershirt, balaclava, jeans, long-johns, woollen socks, woollen mittens, mountaineering boots and snow goggles. I also carried crampons and an ice-axe. We followed a sandy trail of black volcanic ash which sloped upwards at a shallow angle against the main face of El Popo. I found the altitude retarding my progress and much of the time I was near the tail of the group.

After nearly three hours hiking we came to a hut which marked the end of the sandy trail. Here we rested for about an hour. After half an hour a group of around 22 Germans had swelled the hut to bursting point. They were on a mountaineering tour of Mexico and had very sophisticated equipment and loads of food. Just after dawn (around 6am) we left the hut and began the short climb to the snow-line where we put on our crampons. It was very exhilarating with the sun coming up over the edge of the mountain.

At first I climbed quickly and was soon at the front of the field of around 33 climbers. However, after a while my energy began to drain and my pace slowed down considerably. I was still far ahead of the Mexicans but the Germans were overtaking me. One of the Mexicans, Alexandro, caught up to me and we climbed together for the rest of the journey. Around 1pm we reached the crater at an altitude of 17,800' (5,400m). Here I dropped to the snow and drank some water. From where we were we could see the highest point of the crater wall about 400' (120m) above us – a prospect not compatible with the state of my fitness. Altitude sickness was also affecting me so I lay down on some scree on the inner edge of the crater wall for about half an hour. The other Mexicans still had not reached the rim of the crater and appeared to be quite a way off. The inside of the crater dropped down very steeply to a small lake at the bottom. Sulphurous fumes were emanating from various points on the inner surface of the crater walls.

Around this time I decided I needed to find a 'comfort station'. As there was a dearth of Portaloos at the top of this volcano I decided to head down. I told Alexandro I was heading down on my own and set off at a good rate. Several hundred yards down the slope I passed the rest of the Mexicans on their way up. I chatted with one of them for a few moments before continuing on down. On two occasions I fell but managed to stop myself with the ice-axe. My descent followed a somewhat haphazard path until I reached the base of the snow slope. There I went onto a scree slope and continued the descent at a much faster rate. After about 20 minutes I came to the end of this slope and looked around for the path back to Tlamacas. It was nowhere to be seen. Nothing was familiar and nothing was visible. The situation was made worse by the clouds which had moved in on the mountain. On reflection later, I came to the conclusion that I had probably descended too far and was not able to see the path as a result. It soon became evident that I was lost and as the afternoon wore on I realised that I would probably have to spend a night out in the open. After moving laterally for some time, in both directions, I thought I could see a road and parking lot near the edge of the forest. It took about half an hour to descend to this place which turned out to be a mirage; no parking lot, no road, just an area at the edge of the forest.

My only water supply was a Spanish wineskin containing about half a litre of water. At this point I decided to try and follow a stream bed to a larger channel which might contain water and lead to habitation. The first stream bed led to a larger channel which was also dry. I followed this for about an hour, pushing branches away from my face, jumping from boulder to boulder, climbing down small dry cataracts and gradually getting warmer from the exertion and the lower elevation. Eventually the undergrowth forced me to halt and consider finding a way out of this channel. After a while I found a climbable region and managed to haul myself out of the stream bed and back into the forest. By now it was approaching sunset so I sat down and took stock of the situation. Walking through the forest without the aid of a path had proved to be a very slow process; it could take me days to find someone and no chance of anyone finding me. On the other hand, the mountain offered water, in the form of snow, on its upper slopes and if anyone was looking for me they would be searching the mountain not the forest. The only obstacle was getting back up to the snowline. Lack of food and water, the altitude and 16 hours of exhausting climbing the previous day would make the journey to the snowline something of an endurance test. Once up there I knew I could last well over a week as I had a little chocolate and two toffees which could prolong my fast. Furthermore, it would be a lot easier to circumnavigate the mountain at the snowline than at the base and at that height I could spy the countryside for landmarks. So I decided that my future lay with the exhausting climb back up to the snowline. Having made that decision and aided by rising panic about my situation, I thought it best to get a good head start and so immediately dumped my climbing hardware and set off back up through the forest. This bit of mania did not last long, however, as my thirst became too great and I was forced to stop. By this time it was dark and so was not able to find a suitable spot to spend the night. I had also dumped the tube tent with the ice-axe and crampons. Just before I settled myself for sleep I took a sip of water from my wineskin to ease my thirst a little. I gradually calmed down and prepared myself to survive the night. To keep my feet warm I took off my boots and wrapped my balaclava around my feet. My down parka had a hood, which kept my head warm. It wasn't terribly cold but lying on a slope made it difficult to get to sleep. Fortunately the weather was clear with little wind.

At 4:30am I organised myself and commenced the long climb upwards. There was just enough light from the approaching dawn to find my way through the trees and undergrowth. I felt much stronger and found I could manage without water. The rest and sleep must have increased my saliva supply somehow. At any rate the difference was remarkable and by the time the sun rose onto the horizon I was just breaking clear of the forest. Once out of the forest I was able to plan my route up to the snowline. Experience had shown me that trying to ascend a scree slope would be incredibly tiring so I aimed for a ridge (probably a dyke cutting through the flank of the cone) that connected with an iron-stained lava flow just below the snowline. The next few hours saw me steadily working my way up the mountain. I had a sip of water when I reached the ridge and then another further up the ridge. These were the only two times I drank until I reached the snowline. Throughout the climb I tortured myself with visions of reaching the snow and shovelling it into my mouth, then getting as much snow as possible into my wineskin. For that purpose I kept as much water in the bag as possible because the more water I had in there, the easier it would be to dissolve the snow.

By the time I reached the lava flow I was getting pretty tired and was wondering if I would make it by the end of that day. Although it was easy to walk on, the flow seemed to stretch up endlessly and there was a further stretch of scree before I could reach the snow. However, my progress up the flow was easier than I expected and by 2:30pm I was standing at its junction with the final scree slope. From here the lowest arm of the snow field seemed to beckon me upwards and about ten minutes later I sank down next to it and mentally drooled at all that lovely moisture. After scraping off the outer layer of fine ash particles, I gouged out handfuls and crammed my mouth full. The first mouthful of dissolved ice felt wonderful as it coursed down my throat. After repeating the process several times I then filled the wineskin, shaking it vigorously to try and dissolve the snow. To get the snow into the bag I had to take off my mittens so I kept swapping hands, alternately warming one and shoving snow with the other.

Eventually I decided I had enough in the bag and enough in my stomach to keep me going for a while. My stomach was starting to hurt a little from the infusion of very cold liquid. Still I figured it was a small price to pay for getting some fluid into me.

It was now about 3:30pm and the clouds were doing their usual trick of enveloping the mountain. They seemed to move in about 10am and from then until dusk they would keep the mountain in a perpetual envelope, clearing away every now and again, but usually only for a minute or two. The nights would then be clear until the cycle repeated itself the next day. I was very tired after the long climb and seeing a ridge nearby decided to stop there for the night. A stiff wind was blowing so I found a spot on the lee side of the outcrop which afforded me some protection. While I was sitting there staring down at the tree line I noticed a red dot move about 100 yards (90m) above the level of the forest. I stood up, adrenalin pumping and hailed the dot. It seemed to stop and move about in a crazy manner. I couldn't tell if I was hearing my echo or his reply but I felt it was probably the former. During the last 24 hours my eyes had played tricks on me several times and I couldn't be sure this wasn't just another example of that. Having spent the best part of a day struggling up to where I was, I wasn't about to charge all the way down in the hope that it was a searcher. If it wasn't, it would put me back where I started. So I stayed put and watched the clouds roll in and remove my vision. A little while later they parted long enough for me to see that my red dot had disappeared.

Although the wind had dropped quite a bit I soon began to realise that at this height the temperature was going to be a lot colder than the previous night. Consequently I thought it might be a very good idea to get down a bit lower, away from the influence of the snow. Heading in the direction I planned to take the following day, I moved diagonally downhill at a brisk pace. The sun was just setting so I was moving quickly to find suitable shelter before darkness overtook me. There was little danger of hurting myself through a fall as I was travelling across fine-grained scree slopes throughout most of this gradual descent.

Several hundred yards further down I came across another ridge and decided to find a spot against the rock wall to spend the night. There was no wind here but the cold was still intense and I spent most of the night massaging my legs (even wearing thermal underwear my legs were cold). Whenever I stretched out, my body shivered convulsively, forcing me to sit up. This stopped the shivering but made it impossible to sleep. I found I could stretch out lying on my side for short periods so I was able to get a little rest.

From my perch on the mountainside I had an excellent view of the lights of several towns, a long way away and utterly unreachable but still reassuring. I felt incredibly alone up there on the mountain, with the darkness around me, the stars twinkling above me and the cold night air pulsing through me. A companion and a cup of hot chocolate would have been wonderful.

The longest night of my life progressed painfully slowly and eventually the sky took on a pinkish hue with the promise of approaching dawn. I arose just before the sun, organised myself and swore that I would get off this mountain before the sun set again. I waited until the sun was up properly before I set off as I needed to be able to distinguish details in the forest. My plan was to slowly circle the mountain, steadily losing altitude.

This plan worked well and I made very good time. After a while though, it became more and more depressing as I rounded each ridge or large outcrop only to find another wide scree slope on the other side which had to be crossed. This went on for around four hours until I spied the mountain block of Ixtacuahtl, a positive indication that I was approaching the right side of the mountain. Ixtacuahtl was visible from Tlamacos so from that moment on I knew I would get back okay. I imagined that this was how shipwrecked sailors would have felt when first spotting land. A short while later I recognised the face itself. From that point on I began to pick up footprints in the ash, all heading in the direction I was going. At one place I had to make a short walk uphill and I noticed it took me a long time. My energy was dropping again, even with the bolstering effects of recognising landmarks. At the top of this hill I looked down into a very deep valley and an even bigger hill on the other side.

At first there seemed no way across the gorge at the bottom of the valley and even if I did manage to get across the hill on the other side, it looked a good three to four hour climb. I walked down into the valley and almost immediately noticed the sound of rushing water. There were no trees around so the sound had to indicate water in the depths of the gorge. With increasing pace I headed for the gorge, more concerned with trying to see if it was possible to get down to the water than to see if I could find a way onto the other side of the valley and up that big hill. As I approached the lip of the gorge I saw that it was an easy scramble down to the river and that my long thirst was about to end. The river was very small and consisted of several small waterfalls coalescing to form a sizeable stream downriver. The source of the water was very close to where I was, for later on I found that the stream dwindled to a trickle only a few yards upstream. Clearly there were a number of springs feeding the stream in this area.

As soon as I reached the water I took off most of my clothes, emptied the contents of my water bag down my gullet and waded into the water. Pure ecstasy! This was one of the greatest moments I've ever experienced, being able to drink water by the mouthful after three days of no more than a sip at a time, with long periods in between. I refilled the water bag from the stream and drank and drank – I was in heaven. Finally I sat on the bank and ate my toffees and chocolate. The latter was so bad I couldn't even finish it.

When I had dressed and filled the water bag to bursting point I made my way slowly back up the slope. There was now no doubt in my mind that I would make it back alright so I was happy to wander back at a leisurely pace. About 50 yards up the slope I noticed that the gorge rose up and levelled with the surrounding land just upstream from where I was standing. The way to this crossroads was blocked by a large fragment of hill, but I spotted a way around the back of it. In fact there was a sort of path there and in a short period of time I was standing on the other side of the gorge at the base of the last big hill. I drank some more from the stream, had a painful piss (dehydration does that) and then made my way up the hill. The water did wonders for my stamina and so I made rapid progress. By midday I was standing on the top of that hill and looking up at the face I had climbed, complete with path, 48 hours ago. The clouds were obscuring the land at the base of the mountain so I sat down and waited for them to clear.

While I was sitting there I noticed a path just below me, riddled with footprints. I guessed that it would lead me back to Tlamacos so I set off down it at a brisk walk. About five minutes later the sky cleared and I saw my objective a mere 20 minutes walk away, all of it downhill. A feeling of relief, pride (at having found my own way back) joy and a host of other feelings swept over me. Further along the path I spotted my first human being, about quarter of a mile from the path, to whom I waved. There was no acknowledgement but I didn't care. It was over at last and I could see myself dining on gourmet food at the hotel. At about 12:45pm I walked onto the parking area. When I reached the building a guy there started questioning me about my stay on the mountain. He was quite friendly and explained to me that all my belongings were in the hands of the Red Cross which had been looking for me (Cruz Roha Amecameca). He was the driver of a tour bus so he took me up to the TV and radio station at the top of a nearby hill where they announced my safe return to the 'world' (I made the evening news on Mexico City TV). They also gave me a wonderful meal of ham and eggs, fruit salad and tortillas. The fruit salad was the first thing I ate and it tasted out of this world.

After the meal, the bus driver took me back to Tlamacus. On the way, a Red Cross ambulance passed us going towards the radio station so the driver motioned for them to follow us. The Red Cross people were very interested to know what clothes I wore while on the mountain as many people had died of exposure over previous years. They presented me with a list of my belongings, minus the camera, which they said they were holding for me. They told me that Stephen (an American guy I had met before going up the mountain) had been driven to Mexico City by the Red Cross and was awaiting news of me there. They offered to drive me there that night or wait until morning, after a night at their office and hospital in Amecameca. I opted to wait as it would be cheaper. They then drove me to Amecameca, putting on the sirens whenever they passed through a town (they enjoyed it as much as I did!). I was sitting next to the driver feeling like a king. When we reached the town they took me to the shop of the boss of the Red Cross. He gave me two Turkish coffees which I interpreted as a gesture of very definite goodwill. A bit later they let me use the shower facilities in the rear of the shop – hot water and soap – that was another high in my life. Everything one normally takes for granted in normal civilised society takes on enormous significance when one is deprived of them. After my shower they took me to a restaurant and then gave me a bed at their headquarters.

I slept pretty well and awoke to find my face beginning to peel. Numerous small blisters had developed and my whole face was very red. About 10 in the morning one of the guys drove me to Mexico City, a distance of about 40 miles (60km), where I was able to retrieve my gear and meet up with Stephen. The Red Cross then drove the two of us to the Fiesta Palace in the ambulance where we were able to disembark in front of hordes of American tourists. We checked inside and I took Stephen to lunch where we swapped details of the last three days. He told me that my camera had been stolen sometime between the time he went out to look for me and when the Red Cross took over my gear.

He had risen about 9am of the morning I went up the mountain. Taking my camera he walked along the path intending to take some photos of me returning down the mountain. When he reached the hut he couldn't see any trace of us but was able to join up with three Mexicans who lent him part of a set of crampons. They climbed fast and reached the crater sometime in the early afternoon. When he got back to Tlamacos he realized that I was missing so in failing light went back to the small hut to see if I was there. There were some German climbers there so he tried to borrow a flashlight from them in order to search the surrounding area. It took him ages to persuade them to do even that, let alone help in the search. Finding nothing, he returned and got a few hours' sleep before being woken by the Red Cross who questioned him at some length. They decided to call in the Red Cross Alpine Rescue unit rather than the Police as the latter tend to be very expensive. The Red Cross Alpine Unit was a voluntary organisation. They first checked the local villages to see if I had turned up at one of them. If I had been seen, everyone would know. No-one had seen me of course so they returned and started going around and up the mountain. By that time they all thought I was dead but they continued looking anyway, even firing several flares into the air. Stephen then returned to Mexico City the next morning, along with my gear. A few hours later I returned and the party was over.

Just before I left the country I bought some mandarins from a little girl on the street, but I got more than I bargained for when the little girl pelted me with an orange peel. Some people in the crowd started to intervene so the little girl desisted. I interpreted the action as anti-American sentiment which is quite strong in Mexico.

White Island
NEW ZEALAND

Adventure to White Island

1973
New Zealand

1973 was spent in Auckland completing the coursework requirement of a Master's degree in geology. I stayed in Grafton Hall, one of the University residences, where I got to know Andrew and Sam. The highlight of the year was undoubtedly our trip to White Island to explore the island and check out the volcanic activity in the crater. As the island was privately owned by a NZ solicitor, we cleared it with him before we went. There were five of us in the party: myself, Dave, Andrew, Matt and Ron.

We travelled up to Whakatane together and stayed with my Uncle Bob, kipping down in his front verandah. The next day we found that a local fisherman had organised to take us out there. The weather was calm so we bought provisions for a week and loaded everything into the launch. Going over the bar at the entrance to the Whakatane River was interesting. We got through okay but the dinghy, which was being towed behind the launch, was ripped off in the surf. Beyond the bar the water was relatively calm so we motored along at a fast clip. The owner let me take the wheel for a while, which was quite exhilarating.

After about an hour and a half we reached the island and looked around for a safe landing spot. The main beach had quite a big swell so we opted for a rocky site on the SW side. We stepped onto rocks and formed a chain to unload our gear onto the bank. We then took our leave of the fishing boat and carried our gear up a steep slope and into dense pohutukawa scrub. We managed to find a campsite that was flat, although very muddy and riddled with tree roots. We set up two tents, started a campfire and then assessed our handiwork. It was a most depressing looking camp!

That afternoon I put on my mask, flippers and snorkel, grabbed my hand spear and slipped into the gentle swell off the rocks at our landing site. There were several deep red gurnard swimming slowly around the rocks so I started taking pot shots at them. It wasn't long before I had bagged three or four decent sized fish which I dumped onto the bank. I was starting to get cold and there was another volunteer waiting to have a go so I got out and passed the gear over to him. He added one or two to the larder so that night we had quite a feast.

The next morning we set off to explore the island. We were forced to leave behind Uncle Bob's large water container as it was too heavy to manhandle any great distance. However, we did take the carton of beer, divided up amongst the five of us.

Getting through the pohutukawa scrub was hard going as the roots and branches formed a dense tangle at ground level. Most of us had packs but Andrew was carrying a suitcase! He never complained.

Eventually we managed to clear the scrub and emerge into bright sunlight. Above the pohutukawa the slopes were completely free of vegetation (probably killed off by the sulphurous gases coming from the crater) which enabled us to move a lot faster. The surface was compacted ash so it was easy to walk on. We eventually reached the top and, lying down on our stomachs, carefully looked over the edge. We were at the top of a sheer cliff, looking down several hundred feet into the crater at the bottom where steam and other gases were noisily escaping from the bowels of the earth.

After we had taken some photos of the large fumarole and some panaoramas of the main crater we headed down with the intention of trying to find a way into the crater. We eventually gave this up as the slopes leading down to the crater were very steep. We set up a new camp near the shore from where we hoped to be able to get the attention of the fisherman when he returned. The sea was pretty rough so we didn't attempt to catch any more fish with the spear. It stayed rough for the next few days which delayed the return of our fisherman. It also meant our water supply was getting low. We actually tried to make a still out of my snorkel and the billy but it didn't work very well. We were also talking about raiding one of the gannet colonies for food – that probably would have been quite feasible.

Fortunately the fishing boat returned the next day. It took us around to the main crater and let us spend a bit of time exploring the sulphur factory and the fumaroles before heading back.

Above: Andrew carrying his suitcase!   Below: View of the crater floor.

Above and below: Alternate views of the crater floor.

VANUATU
AUSTRALIA

3

Vanuatu

1974

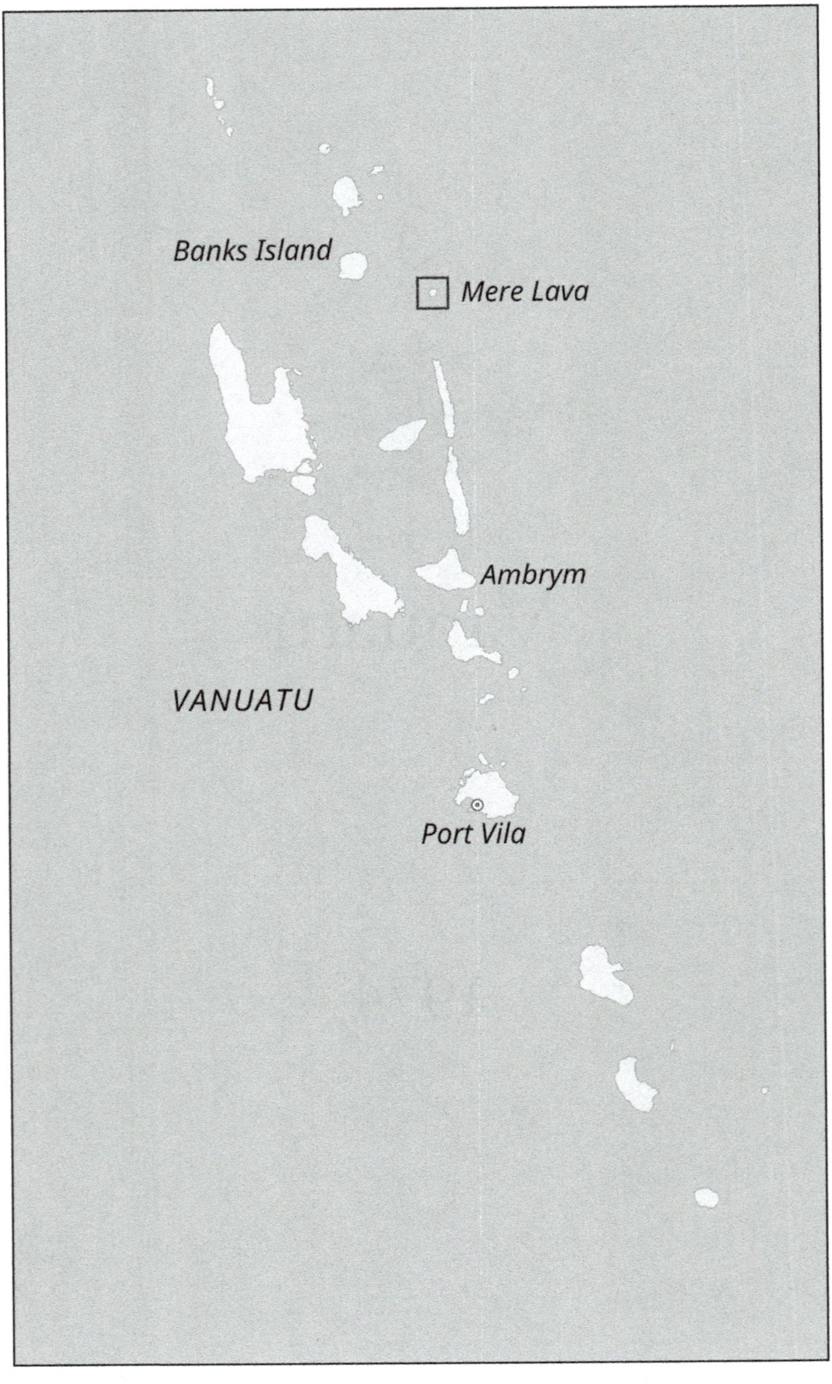
Banks Island
Mere Lava
Ambrym
VANUATU
Port Vila

Part I

8th February 1974
Canberra, Australia

Professor Mason, head of the Auckland University Geology Department, offered me the chance to spend 12 months in the New Hebrides (now Vanuatu) training the locals to become geological assistants at the British-run Geological Survey Office in Vila. As it would give me an interesting area to study for my Master's thesis I jumped at the chance. My salary would be paid for by the New Zealand Volunteer Service Abroad (VSA) organisation.

I departed Canberra at 7:10am on Friday 8th February for Vila via Sydney and Noumea. As I arrived on a Saturday I joined in with one of the VSA organisers who usually spent her Saturday afternoons at the local yacht club. After sailing all afternoon we went to her place for a party with many of the local British expatriates. This was followed the next morning by tennis on

Pango Island at a British club there. I came to the conclusion that the expats had a very strong social life.

On Tuesday I cooked my first meal at the volunteer house. It was a very modest affair of chops, peas and spuds! The volunteer house is located next to the British Gaol, a remarkable institution by any standards. The prisoners would be led outside during the day to cut grass with machetes and the guards didn't appear to carry guns! Even the walls of the gaol were low so that they could easily be climbed. At night the prisoners were locked in cells so we felt reasonably secure living next door to them. On one occasion when the guards had gone back into the gaol for something, the prisoners all ran off to the British Paddock (where they play cricket etc.). The guards didn't seem too concerned; everyone took a very casual attitude to prison life there.

The next day we had a practice for the royal reception in honour of the Queen and some of her family who were due to arrive on Friday. We had been given the task of manning the bars at the reception; serving beer, champagne or whiskey to the guests. When the day came everything went well and I managed to get a look at Princess Anne and her husband. When the reception was over we managed to smuggle out 21 bottles of Pol Roger champagne (the best!) in rubbish bins. As the reception was on Iririki Island we had to carry the loot onto the launch and then take it up to the volunteer house. We had champagne breakfasts for days afterwards! Later we sailed around the Royal Yacht Britannia in one of the yachts from the yacht club.

My first field trip got underway on Wednesday 27th February. We departed on the Lopevi, a small government launch that was ideal for travelling between the islands of the archipelago. We overnighted at Tongoa in the Central Islands before moving on to our first field work destination at Epi Island, also in the Central Islands. On the way we passed the submarine volcano of Karua, its flat nose just poking above the surface of the water. Unfortunately it wasn't active at the time. After Epi we motored past the large island volcano of Lopevi, swam ashore at Paama, then called briefly at Port Vato and Craig Cove on Ambrym. That night we headed for Santo town on the eponymous island, arriving around 8am. I spent the weekend at the volunteer house.

4th March 1974

Our first stop was at Wusi on the west coast of Santo Island, south of the rugged Cumberland Peninsula. We had a bit of an accident with the dinghy in the surf resulting in my losing a shirt and damaging my flashlight. We set up camp on the beach to avoid mosquitoes. The next day we worked our way up the coast, noting the lithologies *in situ* and as float (loose pieces of rock that are not connected to an outcrop). On the 6th, Dean (Chief Geologist) took one of our field assistants (Flynn) to Port Olry on the north-east coat of the island for some medical treatment while I led a traverse up the Pemetch River on the north-east coast of Cumberland Peninsula. We traversed about 2 kms up river, mainly encountering coarse rudites (specifically, poorly stratified, polymictic paraconglomerates) and volcanic breccias. We also encountered interbedded basalt flows and flow breccias and some pillow lavas. On the Thursday we waited at a nearby village for the Lopevi to return. The boat arrived around 2pm without Dean who had taken Flynn to Santo town. So I undertook a quick traverse up a nearby creek with Robbie, the head field assistant, before retiring for the night on the boat. The next day Robbie and I did a more thorough assessment of the geology up the creek and again slept on the boat.

9th March 1974

On Saturday we travelled south to the Piamatsina River where we spent three days detailing the geology and camping beside the river. On the second day our progress up river was blocked by a large waterfall, although there wasn't a huge amount of water coming down. We ascended the left side of the cliff then traversed across the face of the fall to the other side. As I gingerly edged my way across the face, my foot slipped and I started to slide. Luckily for me one of my field assistants, Greg, shoved a pole under my foot and arrested my slide. Everyone got across safely and we continued our ascent of the river. On the third day we were forced to climb a second waterfall before returning to the river mouth and a campsite on the beach. We waited on the beach a further day as the Lopevi had been delayed. The boat arrived in the early afternoon of Wednesday 13th March, in heavy seas. After collecting another geologist (Nelson) en route, we headed for Port Olry, where I slept on board.

The next day I returned to Santo town by taxi with Nelson and Greg, as Greg was ill. Santo is a real wild west town with occasional tribal fights in the street. There were plenty of hotels and restaurants but most of the roads were unpaved. Drinking in bars seems to be one of the main amusements. The next day we flew back to Vila on a trilander, a journey of approximately one hour. The following day started propitiously but ended disastrously. A few of the volunteers and I drove to the Cascades, a series of small waterfalls in a river near Vila that is one of the favourite diversions for Vila residents. We swam in the pools enjoying the cool, swirling water. However, a bit later I was walking beside the river when I tripped and fell into a nangalat bush on the side of the track. The nangalat is a family of stinging nettles that gives a severe sting for at least 24 hours and then a painful affected area for several days. My right arm was badly affected.

The next few days were spent in Vila looking at foraminifera (forams) from one of the calcarenites we collected from the Cumberland Peninsula and doing a lineament analysis of the Cumberland area. Also swam across to Iririki Island, had a dinner party at the house of one of the colonials, and attended a Tongan wedding reception.

8th April 1974

Posted a letter and a present for Dad to my parents and received one back from Dad. I also finalised the itinerary for my second field trip and did some reorganisation in the volunteer house. While I was at work I discovered a javelin in the Geological Survey office, so I took it home to have a practise. Cooked Spaghetti Bolognaise for dinner that night.

On Tuesday, after work, I had my first workout with the javelin, managing to throw it 37 paces. Destroyed a wall in the volunteer house as part of the renovations and also played tennis at the club. On the Saturday there was a cricket match on the British Paddock between the administration and the banks and finance companies. I played for the administration and afterwards we went to the Hotel Rossi for dinner.

15th April to 12th May 1974

Travelled to Hideaway Island in the rescue boat and came back in a yacht called the Janus Lee. On Thursday the 18th I flew to Santo to meet the Lopevi launch, then spent the next 24 days doing field work on the Cumberland Peninsula. Arrived back at 8:20am on Sunday. Spent the day painting a room in the volunteer house, playing tennis and planting spuds.

13th May to 30th June 1974

Apart from a hectic social life over the next few weeks, we did some scuba-diving at Pango, Erakor lagoon and behind Iririki Island, mostly at night. Managed to find several cone shells and the odd cowrie. I also did some tennis coaching over this period. On 14th June I took part in an athletics meeting as part of the Queen's birthday celebrations. I came last in the javelin. Got blotto at the ball that night.

1st to 26th July 1974

My third field trip to North Santo commenced on Monday 1st July. Together with field assistants Robbie, Tom, Greg and Edward, we flew to Santo town and, after purchasing supplies and talking to the district agent regarding the security situation in our areas of interest, loaded up the launch Lopevi and headed off for Big Bay. We overnighted at Crab Island and then proceeded to Cunabo to pick up four porters. From there we travelled to the mouth of the Ilava River for the start of our traverse.

Our plan was to traverse the river to the headwaters, collecting samples of primary and clastic volcanics, as well as detrital and tuffaceous sediments. We also planned to measure the altitude of dykes and strata, current directions, degree of induration of breccias, amount of limestone etc. We split up at the Pepoussa River fork so that Robbie and Edward could map that river while I took the rest of the team up to the upper reaches of the Illava River. We rendezvoused on the 6th of July and then returned to the coast on the 7th to meet the Lopevi, which took us to the Beesel River for our second river traverse. We completed this traverse on the 11th and then proceeded to the Piamaeto River after changing porters. This river mainly hosted greywackes, calcarenites and calcilutites but we found intrusives further upstream. We also examined several other rivers draining the Cumberland Peninsula before returning to Santo Canal on the 25th of July.

27th July to 23rd August 1974

Back in Vila, I spent the next month processing rocks and data from the field trip and made plans for my thesis field work on Merelava Island in the southern Banks Islands.

26th August to 11th September 1974

First Thesis Field Trip

Our plane to Santo was scheduled to leave Vila at mid-day on the 23rd of August, which it did, though without our two field assistants; they were late getting to the airport so Air Melanesia used their seats to accommodate the extra luggage of a pop group. Adam and I and all the luggage made it to Santo safely and we kept our fingers crossed that the boys would make it up here on a subsequent flight before our departure for Sola on Vanua Lava at noon on the 24th. We booked into the Les Castors motel (1000 francs a double - about $9) and then proceeded to pick up outstanding items of equipment and food for the coming trip. Most of our luggage had been sent to Santo on the Lopevi and transferred to the Police boatshed until our arrival there. We had to pack the food for Merelava separately so that it could be put on our transport vessel, the Mangaru, directly in Santo and would be waiting for us when that boat picked us up from Losolava, Gaua. So we spent the afternoon of the 23rd and the morning of the 24th running around sorting things out and making last minute purchases. To further confuse things, the schedule on the Mangaru had been changed so that we had to arrange a new plan with the British District Agent.

To complicate things still further, I got a touch of flu just before leaving Vila so I wasn't able to think very clearly and consequently several items were forgotten – like toilet paper. Several disprins removed my fever and gradually my condition improved (there was no room for illness in our plans so I psyched myself out of it).

At the airport, Air Melanesia informed us that the plane was overloaded, so we were forced to leave one of our field assistants behind (Greg). Even then we only got our packs on at the expense of someone's 40hp outboard motor. The owner of the motor was a guy called Graeme, an archaeology PhD student whom we nick-named Doomsday because he kept forecasting difficulties for us, telling us that his motor was going and that our packs were being offloaded. When we arrived in Sola after a 45-minute flight, Doomsday found that in addition to leaving his motor behind, they had also left his pack and mail. We left Greg with instructions to catch the Wednesday plane to Sola so he could join us for the Gaua and Merelava sections of the trip. Doomsday gallantly offered to transport our luggage from the airstrip to Sola in his boat while we followed on foot. The walk only lasted about 20 minutes and we were soon ensconced in one of the huts belonging to the Van Nerum 'hotel' complex at Sola. Van Nerum is the 'king' of Sola and is in the process of building a group of tourist cabins on the water's edge. The only one with a shower and stove was the one he uses when he makes his infrequent trips to Sola, so we managed to have that made available to us for our stay. We weren't charged any more, even though we stayed several nights. He operates a store there which sells mainly tinned food for the locals, as well as cold beer (Castlemaine XXXX), sweet biscuits, some clothing items etc.

The next morning we managed to get hold of a dingy with a

very small seagull outboard motor and persuaded the owner to let us keep it for a few days, though at some considerable cost. The boat cruised along at walking pace and then just near Point Patterson the engine conked out and wouldn't start again. So we paddled into the beach and continued our progress on foot. We were heading for the village of Lalngetak, about two-thirds of the way up the east coast of Vanua Lava. The walk there took us about two hours and led us through mangrove swamps, coastal bush and along beaches of coral sand. At a small village near Lalngetak we bought some shells from the villagers and tried to hire a guide to take us up the mountain to the solfataras. The only available individual demanded eight dollars a day so we politely told him where to get off. Robbie, our top field assistant, had made the trip a few times so we relied on him to find the path for us.

We arrived at Lalngetak about midday and stopped there for lunch. The village consisted of about four huts, one of which was owned by a one-armed Frenchman, a fantastic character who spoke no English and communicated with us in a mixture of French and Bislama (the Vanuatu pidgin English). We would often start a sentence in French and then finish it off in Bislama (me) or New Guinea pidgin (Adam). Jacky, as he was called, had another Frenchman staying in the hut with him who also spoke no English so we had a rather hilarious time. Jacky had outfitted the hut with a kerosene fridge (containing cold Fosters), abundant French delicacies, portable stove, Coleman lantern, carpet, bed etc. and was obviously very comfortable there. His hospitality was impeccable and before setting off on our climb up the mountain we were treated to French wine, cheese, bread, cold beer and sauerkraut. We were then in no ideal shape to lug a heavy pack up a steep slope on a scarcely recognisable track, but we did.

Quite a bit of the track was overgrown so Robbie had to hack a path for us with his machete. The track passed quite close to Sulphur River which drains a wide area of solfataras and hot acid springs. After a couple of hours we made camp beside a cold freshwater creek about 15 minutes' walk from the solfataras. As soon as we arrived, Robbie started to organise the setting up of camp, which left Adam and I free to go off and have a bath in the stream – a wonderful luxury after a day walking through the bush. By the time we had finished our bath, Robbie and Greg and the porters had finished setting up camp and had boiled some water for tea. Later we inflated our lilos, ate some rice and mackerel and retired to bed around 8pm. There were eight of us altogether, sharing one fly-sheet, so we were packed in a bit like sardines. In a conversation with a French woman doctor some time later, she professed amazement at our sleeping at close quarters with the local people. Her attitude represents a common fallacy amongst casual observers, that the locals are dirty. In fact they are scrupulously clean and wash themselves and their clothes every day, which is more than can be said for most Europeans (including myself!). In the town, the locals are always dressed tidily and are well groomed.

We got going about 7:30 the next morning and walked to the Sulphur River where we made temperature and pH measurements and estimated flow rates. The highest temperature recorded in the river (measured in mid-stream to avoid local effects caused by peripheral hot springs) was 57°C. The lowest temperature, recorded above the junction with the cold stream mentioned earlier, was 30°C. The pH (acidity) was in the range 1-2, which is at the extreme acid end of the scale. The river was therefore a hot, dilute solution of sulphuric acid. Needless to say, we hopped across stream boulders with considerably more care than usual. Chemical analysis of the acid sulphate waters, carried out on our return, indicated high levels of chlorine and fluorine, which indicate a degassing magma at intermediate depth.

Some of the hot springs feeding the main river had temperatures as high as 87°C. Thin layers of sulphur had been deposited on the bottoms of the hot pools and along the sides of the main river, and many of the stream boulders displayed extensive alteration effects due to the action of the hot acid. We collected a few samples and then moved on and up to the solfataras, which occupied several hillsides above the river. The main feature of the solfataras area was the numerous small sublimating sulphur fumaroles (gas vents) characterised by bright yellow deposits of sulphur around the orifices. The temperature of the escaping vapours inside these vents was 98°C. There were two large steam-gas fumaroles in this area not sublimating sulphur (due to lower temperatures) and one of these provided us with some amusement for a while; we tried to block the vent with large boulders so that we could cause a small explosion to clear it. However, the pressure of the escaping gas was too strong to successfully block the vent and we weren't able to achieve our desired aim. During my wanderings around the area, one of my feet fell through the edge of a hot pool, but I was able to throw my weight onto the other foot and only sustained a partial immersion. There was also a boiling mud pool, similar to the ones at Rotorua in NZ.

We spent some time looking around this area and then followed a track further up the mountain to another hot spring region just below the summit of Mt Suretamati. The whole complex is considered a dying volcano in the solfatara stage (where no further eruptions are predicted and sulfurous gases and steam are emitted). The walk to the second area took about two hours and the vista wasn't as spectacular as the lower area. There was one large pool in a constant state of fountaining. The muddy water was being ejected to a height of about two metres and was accompanied by the emission of a considerable amount of steam and sulphurous gases. Not far away was the site of a tragedy some years ago when a Frenchman fell into a large hot pool and was killed. The pool wasn't immediately obvious and I wasn't about to explore the area too closely in case I found it the same way as the Frenchman.

It had been steadily raining for most of the day and by the time we had examined the second thermal area we were pretty miserable, so we headed back to camp. While trying to measure the temperature and flow rate of one hot pool beside the main river, my foot slipped into the water up to my ankle, but apart from being uncomfortably hot there were no deleterious effects. The temperature of that pool was 76°C. No further mishaps occurred and we eventually arrived back at camp at about 3pm. Our campsite had become a quagmire so we decided that rather than spend an uncomfortable night there and return to Lalngetak in the morning, we would pack everything then and there and proceed directly to the village that afternoon. This we did and it was fortunate as the rain came down in torrents and would have flooded out our camp had we still been up on the mountain. When we reached Lalngetak, Jacky gave us some French 'grog' (Scotch, hot water and sugar) and then served up a fantastic meal of locally caught lobsters and more French wine. Apparently lobsters (probably crayfish) abound in these waters. That night we slept the sleep of the exhausted.

The next morning (27th) we had breakfast with Jacky and Andree and then walked a short distance to the River of Crocodiles which, as the name implies, contains a few of those saurians although we weren't fortunate enough to see any. Jacky reckoned a couple of 2.5m crocs had recently killed a dog at the river. They must live somewhere up on the riverbank as there is almost no water in the river. We walked slowly upstream sampling every different volcanic rock type we could find, irrespective of whether it was float or *in situ,* the idea being that the wider the range of composition, the more representative the collection and thus the more credible the interpretation placed on the analytical data we would later obtain from these rocks.

After lunch I set off with Robbie and one porter to collect from a stream draining Toovalava, a peak of older volcanics. A short way upstream a waterfall formed by a resistant lava flow capping underlying pyroclastics entertained us for a couple of hours, firstly trying to climb on to it and then, once up there, trying to collect a fresh sample from the flow. We eventually got some half-decent samples and returned to the main river. Adam was still suffering from a headache and it was again pouring with rain so we then headed back to Lalngetak. Meanwhile a plane had just landed and we saw Greg walking along the road with a Coleman lantern and a map we had forgotten. This meant that we were now fully equipped for the rest of the trip. By this time it was early afternoon so we decided not to try anything further that day and made plans for the next two days' work. Van Nerum's manager opened Van Nerum's hut for us so we had a shower and cleaned ourselves up and then relaxed with a beer afterwards. It was interesting to note that there were no resident English-speaking Europeans in Sola and possibly the whole Banks Group. There is a French school at Sola and a French radio station, but little else.

The next morning, Thursday the 29th, we set out for the Befornonom River a couple of miles past Mosina village on south Vanua Lava, the aim being to sample float coming down this river and so get a representative collection of rocks draining Tolav – a cone of young volcanics. The next day we retraced out steps and continued on to the Peddol River just past Crepeta village. The walk there took 3.5 hours so it was with considerable pleasure that we viewed an excellent spot for swimming and cooling off after our arduous trek. Where the track crossed the river was a lava flow, which we sampled. On the way back we stopped for a beer with one of the locals in Mosina village, not really a welcome refreshment as it was warm. The guy was somehow responsible to the British administration for Vanua Lava such that he was their representative and would keep abreast of anything interesting happening on the island. He had spent some time in the Solomons and spoke quite reasonable English. We eventually arrived back in Sola about 5pm and immediately bought some COLD beer from the store. The beer up there was all Castlemaine XXXX.

Finally we boarded the Mangaru again and set off for Merelava.

1 mile = 1,760 yards

Merelava Field Work

Merelava volcano has been inactive since 1606. What was this island doing here? The main volcanic chain of the Vanuatu archipelago lies 80kms to the west. The only near neighbour was the hat-shaped Merig Island, 20kms to the west. If the volcanic chain represented the interaction between two colliding crustal plates then what was happening here 80kms behind the zone of interaction? This question had aroused my initial interest in the island, but published reconnaissance reports on the rock types occurring there convinced me of the need to visit the island. Fragments of deep-seated rocks called peridotite had been carried up by rising molten rock and were present in some of the lavas. Such fragments are extremely rare in island arc environments and indicated some unusual happenings beneath the volcano. The answers to these questions would be the focus of my post-graduate research over the next few years.

The Mangaru has four berths for Captain and passengers, the crew sleeping up front and down below somewhere. We found only one spare bunk so I *volunteered* to put a lilo on the floor. Greg made himself comfortable on a bench at the back of the boat and Robbie eventually climbed into the lifeboat and curled up there. Twenty minutes after leaving Sola the first symptoms arrived and ten minutes later I was leaning over the rail. This time however, one good "yawn" had no alleviating effect and I spent the rest of the night intermittingly clutching the rail and trying to get a few minutes rest on the lilo before the next impulse came. By the time the sun had risen above the horizon and added a new dimension to the black hulking shadow of Merelava I was pretty shot, and was in no mood to appreciate the aesthetic side of the vista.

Adam and the boys hadn't slept very well either so our field work on the island promised to get off to a slow start. The Mangaru dropped anchor off a small lava flow promontory at Lequel village and preparations for breakfast were quickly organised. Apart from ourselves and the education officer who had to spend a couple of hours on the island to pay salaries, there were two other passengers disembarking; one was a young New Hebridean lass about to take up a teaching position at the English school at Lequel, whereas the other was just returning home after a brief sojourn elsewhere in the islands. I declined the offer of breakfast and watched Adam devouring bacon and eggs, fruit juice, rolls and jam etc. with considerable anguish – it would be some time before I would have the opportunity to eat another meal like that. The cook was a real clot; he had run out of corn flakes so had bought a large packet of Smiths crisps in its place, which he placed on the table in front of everyone's cereal bowls. After some appropriate prompting he eventually found a suitable replacement.

As soon as breakfast was over we climbed down into the lifeboat and proceeded to shore through a fairly heavy swell. By this time all the villagers from Lequel village had come down to the rocks to help us offload and step ashore. They formed a line from the sea to a point further up the slope and passed all our luggage up to safety in this manner. They then gave us a hand to step onto the rocks. There are no beaches *per se* on Merelava, the closest approximation being a short strip of bouldery shoreline at Tasmat St Stephen village, the quietest and only real anchorage on the island. The village chiefs offered us the resthouse for our accommodation and the local people carried our equipment up the path there without any fuss or hassle. The chief said no Europeans ever stayed on the island (except the English volunteers who had built the resthouse) so they were quite thrilled to have us around for a few days. The resthouse was built originally for the purpose of attracting government people to stay there but it has apparently had no effect.

From the coast, the island has a rather fearsome aspect. The slopes of the volcano rise straight up from sea level to 3,300′ (1,000m) at an angle of between 30-35 degrees. We worked out that there must be three classes of people on the island: those with their left leg shorter than their right who walk around the island in an anti-clockwise direction, those with the opposite symmetry who walk around the island in a clockwise direction and those with both legs the same length who walk up or down the mountain. When a clockwise walker marries an anti-clockwise walker, the offspring have equal length legs and only walk up or down. Actually, the island has an excellent network of wide paths which makes walking a relatively pleasant pastime.

The house had concrete foundations and walls, New Hebridean windows (boards, kept open with poles), a tin roof, a tank at the back and an enclosed pit toilet. The windows are a common feature throughout the islands as glass would be hazardous during cyclones. The people said there had been no rain for over two months and all the tanks on the island were dry. There was still a little water in the crater lake and they were also obtaining a small amount of water from the trunks of some trees. We had come prepared for this and had brought about 12 gallons with us from Santo. Upon arriving at the hut after the five-minute walk from the shore, we put on a kettle and invited the three chiefs to join us for a cup of tea. Our work being dependent on their cooperation and goodwill, we were careful not to abuse their hospitality. They spoke Bislama so we had no trouble conversing with them. When Adam or I couldn't understand, Robbie or Greg would translate for us. Later we talked to the head teacher of Lequel school and he told us there were over 1000 people living on the island. Considering the area of the island is 16sq kms, that is quite a high population density. There are five villages on the island: Tasmat St Stephen, Lequel St Paul, Lewitnok St John, Auta St Barnabas and Levetmissi St Luke. Each has a church and a local Anglican minister. There is a very modern French school at Tasmat with a well-organised rain-water collection system. Lequel has an English school and is building a larger school to accommodate the increasing population. Tasmat appears to be the largest village. The French Co-ops department operates a few small stores at these villages, selling such items as sugar, canned meat and fish, Sao biscuits, rice, soap, articles of clothing, toiletries etc. We found out only after finishing in that area that there was a store selling cold beer at Levetmissi. Why they sold it there and not at Lequel or Tasmat we couldn't guess. Possibly more ships call there than elsewhere, though it is one of the most exposed parts of the island.

Most of the villages are concentrated around the northern part of the island and are spread over the slopes up to a height of about 150m. All ground has to be terraced before houses can be built and crops laid out. The villagers build low stone retaining walls to prevent erosion of the terraces and commonly the houses are built on a stone wall framework. Several types of architecture are practised and an array of building materials is used. Walls are sometimes made of interwoven bamboo, planking or stones.

Surprisingly, much of the roofing is made of local materials rather than corrugated iron, which would be more practical for collecting rain water. Agriculture is the slash-and-burn variety, which is widespread throughout the south-western Pacific. The people cut down the vegetation on a hill slope then burn it (which removes all the humus but does add nitrogen and potassium to the soil) and terrace the ground. Often these slopes lie directly above huts which have been built on a terrace further down. One wonders what happens during a rainstorm after the vegetation has been cleared off a 30 degree slope. We noticed several instances of land slippage breaking through retaining walls and in places threatening the odd house or two. You would think that 1000 plus people would be able to construct a more permanent and practical method of erosion control and land use but they seem to have no inclination to improve the situation. The children just copy the methods of their parents and grandparents and so it goes on. The same can be said of the water situation there. Every village should have a large water collection and storage system that would carry them through dry periods. Luckily there is no shortage of coconut trees on the island.

Notwithstanding the somewhat precarious living conditions, the people are extremely happy and their society as a whole is geared to successful communal living. There is no inter-village friction and within each village everyone lends a hand when somebody wants to terrace a bit of land and build a house. The farming appears to be communal too. The system works well in times of disaster as happened in 1957. A violent cyclone destroyed all the crops on the island and people were forced to find food in as unlikely places as the hearts of tree ferns. All the people worked together, pooling their food, and after about two years finally got back to normal. Although these people had no water for washing, most of them kept clean by swimming in the sea every day. They also wash their clothes in the sea. Until it rained (and even after this) we made a point of doing this too – a dip in the ocean after a hard day scrambling over the broken surfaces of lava flows was sheer pleasure.

But, to get back to the beginning, by the time we had finished our chat with the chiefs and organised ourselves I was feeling particularly crummy, so decided not to try and bust myself doing field work in that state and retired to bed (lilo on the floor). Adam took Greg with him and spent about three hours working along the coast west from Lequel. Robbie wasn't needed so he retired to bed too. The others came back around 1pm and, after a light lunch, everyone retired to bed again for the duration of the afternoon. In the evening we cooked our rice on an open fire and added tinned pork and mackerel for our dinner. We had a few tins of fruit, a packet of biscuits, a couple of tins of orange crystals and even some peanuts to relieve the mealtime monotony. The highlight of most of my bush meals, once my appetite was sated, was to sip a large mug of milo, and if I could sit on a chair while I did this, the feeling of wellbeing was complete. The one item we forgot to take was a bottle of tomato sauce. Of course another item we forgot, yet which we miraculously got by with considerable luck, was toilet paper. On Vanua Lava we found half a roll in Van Nerum's hut, and in the house on Merelava there was another half roll. A bush bog isn't so unpleasant when its dry, but when it's raining and all the leaves are wet...

The next morning the weather looked promising so we decided to climb up to the crater and do our work there while the sky was relatively clear. We also wanted to get the hardest day over with early in the piece. We recruited a guide from the village and set off up the mountain. At first the going was easy, but after we had passed the 150m mark and had left the highest terrace level behind, the path quickly degenerated into an overgrown, barely recognisable route. The path was now heading almost straight uphill at an angle of about 35 degrees, which increased sharply in places, causing us to climb with the aid of tree roots and vines. We had our first rest half an hour after setting off, and then at more frequent intervals after that. The vegetation was now becoming more stunted, tree ferns were becoming prolific and the undergrowth seemed denser. We carried a full half-gallon water bottle with us for the journey up and another empty one for filling at the small lake in the crater to supplement our supply at base. The tree ferns slowly gave way to dense grasses and then virtually to bare rock near the top. Walking up the rock slope, although steep, was a pleasure after two hours of stumbling through undergrowth lower down. The wind was steadily increasing as we approached the crater rim and by the time we had reached it the wind was gusting to around 30 knots (56kph). This, combined with the light drizzle and cloud we had entered a little way below the rim, served to make our stay at the top rather chilly. Still, it was a relief to cool off from the sweat-soaking hike up the hill. I measured the altitude at that point on the crater rim to be 3,305′ (1,007m), uncorrected for barometric fluctuation.

We didn't dally there but continued along the rim for a while, then climbed down into the crater via another overgrown path. Visibility at the top of the mountain and in the crater was rarely more than a few yards (as we were in cloud) so we were not able to get any sort of perspective of the crater and the location of the small cinder cone inside it. The path inside the crater descended through moss forest, largely composed of tree ferns heavily mantled with moss and epiphytes of various sorts.

Moss forest always conveys to me a sense of eeriness, almost as if you expect some strange prehistoric creature to materialise through the misty gloom. The stillness and dampness pervades everything to the point where time seems to stand still and a wristwatch seems a gross incongruity with the magical aura present. One feels one should be still and absorb the tranquillity of the scene; to march through looking for rocks to smash seems the height of blasphemy in such surroundings.

The water depth in the lake was no more than a few inches, just enough to submerge the water bottles without stirring up too much silt. We decided against trying to climb Star Peak, the cinder cone inside the crater, as the vegetation cover looked too dense to sample properly and certainly wouldn't warrant the effort in climbing the cone. The coolness of the air at this elevation was starting to cause some discomfort to our scantily clad guide and porters, a situation that presaged serious consequences if we remained here too long. On an earlier visit to another volcano in this archipelago I had witnessed several islanders suffer an attack of malaria brought about by exposure to low temperatures. Chloroquine tablets had solved that problem but I wasn't keen to experience a repeat performance. Diurnal temperature variations at sea-level in these latitudes are limited, so the islanders have little tolerance to excessive falls in temperature.

After a quick lunch beside the lake we retraced our steps, collecting a few samples of volcanic bombs and ash on the way. We descended the mountain by a different route, which seemed like an eternity of undergrowth and clinging vines and eventually arrived at the village of Levetmissi about 15 minutes' walk to the west of Lequel. It was still relatively early in the afternoon but we were pretty tired and decided to call it a day. I paid the guide and another small boy who had tagged along with us the standard rate of $3 a day. The government had permitted me an imprest of $150 in addition to our air fares up and back. Adam had obtained a similar amount via a grant from Auckland Uni. As it so happened, we spent just about every penny of that money. The small boy became quite attached to us as his parents had moved to Gaua several years ago and he was being brought up by his grandparents. He was only ten but could shin up a coconut tree in a matter of seconds and he could walk the legs off either Adam or myself. He used to bring us native oranges (rather like a pomelo), coconuts and a kind of nut which was very abundant on the island.

About 4pm that day we walked down to the coast below Lequel village and had our first swim in the sea. There was a small rock stack just beneath the surface in the middle of a small bay, so we dived in and swam out to it, finding that we could stand on it comfortably with our head and shoulders out of the water. We spent the next hour relaxing in the water and were joined by several younger members of the village. The locals thoroughly enjoyed their daily swim and kept up a noisy banter. At this time of the year the evenings were still cool and our only regret was the absence of twilight. As it was, we still sat outside for a while after dinner enjoying the sound of the sea and the coolness of the night.

The next morning we headed eastwards walking along the coast sampling lavas at intervals. The coastal cliffs revealed magnificent layered sequences of lava flows and breccias, representing a seemingly endless series of eruption cycles. These cycles always started with a violent phase, as gas pressure building up over long periods of time finally overcame the confining pressure within the volcano and opened a passage to the surface. The initial gas-rich eruption pulverised the solidified lava in and around the vent and hurled the resulting volcanic ash high into the air, along with particles and fragments of *partially* solidified lava. The accumulation of all this fragmental material formed the breccia deposits so strikingly displayed in the coastal cliffs. As the gas pressure in the magma chamber gradually subsided, subsequent eruptions became less violent and produced lava fountains which fed rivers of molten rock, consuming everything in their path as they plunged down the flanks of the volcano on their way to the sea. Later eruptions buried the products of earlier eruptions and over thousands of years the volcanic cone evolved into an awe-inspiring edifice, 1000m high.

We stopped for lunch near some coconut trees and dined on the usual fare of tinned mackerel, cabin biscuits and coconuts. Green coconuts contain a slightly effervescent and sweet liquid that is an excellent thirst quencher. The Merelavans never seemed to show any evidence of being thirsty; even the guides who took us to the top of the crater showed no inclination to drink anything during that whole climb. Their adaptation to the water shortage situation was quite pronounced. After lunch we continued walking along the coast examining the flows and collecting samples of anything interesting, until about 3pm when we returned to base. We had another swim, then returned for another gourmet meal of mackerel and rice before hitting the sack at the usual time of 8:30pm.

The next morning we decided to continue where we had left off the day before, walking along the path to our last recorded stop then traversing the coast for as long as we deemed it expedient. The weather was excellent and the coastline very picturesque, with its protruding lava flows and crystal clear water. Hopping across the lava flows was exhilarating and by lunchtime we had covered quite a lot of ground and worked up a good appetite. After lunch we continued walking along the south-east coast for a little way, before climbing up to the path and continuing in the same direction at a much faster pace. As we were looking for a way up, Robbie made a fantastic find; enclosed within an olivine and clinopyroxene-phyric basalt flow was a huge peridotite nodule measuring 10.25″ by 9.5″ (26 by 24cm). The nodule was composed entirely of olivine and clinopyroxene and was fairly obviously a cumulate from the base of the earth's crust. By working out the exact composition of the constituent minerals, estimating the oxygen fugacity of the iron-titanium oxide equilibrium and the temperature of crystallization and by applying certain thermodynamic equations, we could deduce the pressure and temperature of equilibration between the mineralogy of this nodule and the enclosing magma. From the pressure data, the depth of equilibration can easily be derived, a major factor in determining the genesis of the Merelava magmas. Such nodules are extremely rare in island arc environments. In addition to olivine-clinopyroxene nodules we also found recrystallised dunite (pure olivine) and pyroxene megacrysts. The large nodule was made up of small crystals so we collected a full sample bag for microprobe analysis in Auckland.

After this diversion we continued with our search for a way up to the main path. Robbie had managed to climb a vertical 4.5m bank with the aid of pandanus palm roots, a route which we then followed. This required a good jump at the bottom, grabbing the lowest (and very thin) root with one hand, hauling up on it gingerly until within reach of another, much larger root which could then be climbed more easily up to a point where a lateral traverse across several other roots had to be made, finally leading up to flat ground on top of the cliff. Above this cliff was a gentle grass slope, apparently part of a small copra plantation in which several pigs were foraging. We stopped there briefly to drink a few coconuts then walked up to the path. We sampled outcrops along the path periodically until we reached Tasmat when we ceased geologising and just walked on to Lequel. This traverse had taken us the complete way around the island so we were pretty tired by the time we got back, which was around 6pm. There was just enough time to have a quick swim before dinner.

While we were cooking dinner one of the people from the village told us there was a dance on that night at the school hall to which everyone was invited. They even had a local guitar band. So after dinner we made our way to the school where we inspected the setup. We were one of the first to arrive so we hung our Coleman lantern in the centre of the hall and sat down on a bench to watch the band. There were three in the group and they weren't too bad, except that every song had the same basic tune running through it. This is typical of New Hebridean groups – there is very little variety in their music. It was some time before people started to arrive, so by the time the hall was nearly full I was ready for bed. However, I managed to get in a few dances with some of the local lasses before retiring. Some of the young men were very well dressed, which rather put us to shame – I was wearing flip-flops, shorts and a coloured singlet – my best outfit!

The next night we were visited by the village minister who, after much subtle hinting, finally invited us to have some coffee with him at his house; I must point out that we weren't too keen, being pretty tired and hoping to spend our last night organising the packing. This was why we didn't invite him to have coffee with us, apart from the fact that it would have necessitated making another fire and waiting hours for the water to boil. However, he and his family were very friendly and we quite enjoyed our evening there. He didn't discuss religion at all, much to our relief. He said there is virtually no industry on the island as the limited and steep land area precludes the mainstream industries of copra, cattle, timber and fishing. However, the villagers are quite self-sufficient, relying on vegetable gardens, fruit trees, fishing and a few pigs and chickens for sustenance. Daily pastimes centre on development and maintenance of vegetable gardens, construction of new houses, fishing and plenty of tok-tok (talk).

The staple preparation here, as everywhere in the archipelago, is lap-lap, a mixture of grated coconut and one or more of the root vegetables (e.g. taro, manioc or sweet potato) baked in banana leaves over hot coals. The result looks rather like a blanc-mange that has been left to age for a few weeks, but is actually quite palatable. It can be a dessert or a savoury main course, depending on what has been added to the basic recipe. The introduction of regular, if infrequent, visits by trading vessels has meant the inevitable appearance of tinned meat and fish, soft drinks and alcohol and the slow corruption of the traditional lifestyle. Fortunately this island is outside the main inter-island shipping lanes and therefore the traditional patters of life will take longer to succumb to the ways and excesses of modern cultures.

Children are given basic education at mission schools but older children who want to continue their education are forced to board at secondary schools in Santo, the regional centre. On Merelava, the schools teach both English and Bislama. Bislama has some beautiful expressions, like the translation of brassiere – 'basket blong titi'. If your truck won't start, it's – 'trak ia emi bagarap'.

Early the next morning we packed up everything and waited for the Mangaru to arrive. Around 8am she appeared on the horizon and about half an hour later was steaming in to the anchorage. Villagers appeared from all corners and without any prompting, carried our luggage down to the landing site. We were sorry to say goodbye to the island and the people; we had a very enjoyable and memorable stay on the island despite some doomcasters who had disparaged the place so strongly. It is easy to stand on a ship, look up at the steep slopes and think what a Hell of a place it is – and not give a thought to the people who live there, the beauty of the coastline, the peace and health that exists there and a host of little things that make the place stay in one's memory for a long time. The education officer who spent a couple of hours on the island epitomises those who spend as little time on the island as possible and never get to know the people. Much of the government here, both British and French, appear to be staffed with such people. We were near the end of our allotted time in this secluded little world, acutely aware of a community attempting to maintain its identity and integrity against the inevitable and irresistible pressure of change. Would our young guide embrace the 20th century and work in Port Vila or Santo, or would he settle on the island and attempt to preserve the old values and ways? Our memory of a romantic island clinging to a different time would endure, but for the islanders, many would see slow, irreversible change.

The whole village waved goodbye to us as we pulled away from the coast headed directly for Santo. The weather was clear and the sea relatively calm and after taking two nautamine tablets I had a very comfortable ride back. The trip took about 11 hours and was uneventful and a little dull compared to the previous five days.

18th October 1974

Ambrym Island Field Trip

The Geological Survey boat Lopevi departed Vila about 10am and sailed in good weather to Tongoa island in the Central Islands, arriving there at approximately 8:30pm. The two crew members, my three field assistants and I slept on board. Early the next morning we headed for Ambrym via the barely emergent submarine volcano of Karua and the towering pinnacle of Lopevi, the latter quietly emitting a small amount of steam from its NW crater. A radio message delivered to the boat the day before, and confirmed by the evening radio messages, requested us to investigate the state of volcanic activity on the island. Air Melanesia pilots flying over the island had observed an increase in the activity of Marum and notified the Geological Survey.

Ambrym has a very large, flat-floored, ash-covered caldera, 13kms wide and about 600m above sea level, in which lie two active cinder cones, Marum and Benbow, and a third subsidiary cone adjacent to Marum. The caldera stage of the evolution of a volcano involves collapse of the main cone structure into the partially emptied magma chamber immediately below it. This gravitational collapse usually produces a somewhat circular, flat-floored depression with steep walls which may become filled with water to form a crater lake.

The captain of the Lopevi took us around to Ranon village on north Ambrym where a relatively easy passage could be made over the rim and down onto the floor of the caldera within a few kilometres of the active cones. The two cones lie very close to one another and stand about 305m above the caldera floor at their highest points. An Australian/New Hebridean chap living at Ranon told us that the local villagers would not carry for us on the following day as it would be a Sunday, so rather than hang around for a day doing nothing I decided we could haul enough stuff up ourselves to keep us comfortable in the crater for a couple of nights. So, on Sunday morning, 21st October, four of us headed up the mountain heavily laden with bulging packs. On the way we passed through a Nagriamel village, but surprisingly nobody tried to stop us. My experience with Nagriamel natives (those calling for a focus on the traditional, village-centred way of life) on South Ambrym earlier in the year, at Port Vato, showed what they can be like if they decide to be stroppy - being surrounded by 50 natives, all armed with machetes. Anyway we were not hindered this time and after hauling up the gradual incline for about 2.5 hours we found ourselves standing on the rim (still surrounded by thick vegetation) with a reasonably good view of the two craters. The floor of the crater was completely ash covered and vegetation was restricted to clumps of dwarfed trees and small bushes. Marum rises very majestically as a rather lop-sided cone of pyroclastic ejecta, completely devoid of any vegetation and heavily dissected by runoff on its lower levels. Benbow is a lower cone slightly to the west of Marum but little detail could be seen of the structure from our observation point.

The descent to the floor of the caldera took only a couple of minutes and we soon had a camp set up near the crater wall. We pitched the tent here and not closer to the craters because of the former's proximity to a cold spring, water being a scarce commodity in the caldera. The spring issued from a series of bedded tuffs in a small cliff exposure somewhat below the level of our camp. To get down to this spring we had to scramble a long dirty detour to one side of the cliff, an unpleasant journey after bathing in the spring. So I asked one of our troop to see if he could construct a ladder from our level down to the level of the spring, which he accomplished very satisfactorily that afternoon. The only trouble was, he placed the rungs so far apart that each step became a contortion, considerably complicated when carrying toilet articles or water containers in one's hands.

After a quick cup of tea, Robbie, David and I set off across the artificial peneplain of the caldera floor. It was dead flat, sporadically covered with isolated clumps of stunted bushes and the dead ghosts of former trees, the victims of countless ash eruptions over the crater floor. Volcanic bombs of all shapes and sizes, some moulded into aerodynamically perfect shapes, lay scattered around everywhere. I was wearing boots, Robbie was in flip-flops and

David was barefoot during this trek. Volcanic ejecta is typically merciless on footwear so what it must be like on bare feet I can only guess. The walk to the base of Marum took 2.5 hours so the distance would probably have been about six miles (10km) as we were making fairly rapid progress.

As we approached the cone we could see that the lower slopes were heavily gullied and sculptured. With the material being so loosely compacted, it was easily eroded by wind and rain. Robbie had made this climb once before, several years ago, so he steered us on a wide detour to the lowest part of the cone, being about the only accessible route. The 1966 Australian Universities Expedition to Ambrym had discovered that trying to climb the steep, loose and crumbling slopes of the cone was impossible on the steeper side, so had found a more navigable route on the shallow slopes to the north. Our first attempts that afternoon ended in failure as we found ourselves in bottleneck gullies time and again. The landscape had changed sufficiently since Robbie's former visit to completely throw his bearings. With the hot afternoon sun bearing down on our backs, the job became a bit of an endurance test and finally I had to call a halt as my boots were killing my feet (lack of water also contributed). David also stopped and we let Robbie continue the reconnaissance to see if he could find a route up to the rim. A couple of hours later we saw him as a small speck outlined against the rim of the crater. David and I then started our trek back to camp where I gratefully removed my boots and sank another cup of hot tea. Robbie returned about an hour later but could not tell us anything about the state of the crater as he had been too afraid to go too near the edge. We all climbed down to the spring and had a good bath. The water was beautifully cold and clear and felt absolutely marvellous.

That night the heavens opened up and continued throughout the next day. I was determined to get up to the crater rim so all four of us set off in the rain early in the morning and quickly crossed the ash plain to the foot of the cone. Here Robbie took us on a wider detour than previously taken and we found the going a lot easier, although it still involved considerable scrambling. The cloud base had settled quite low on the cone so the latter part of the climb found us walking in very reduced visibility. The roaring sound we had heard emanating from Marum crater the day before was absent now, so I took it to indicate a decrease in the amount and rate of steam emission. The final slope to the rim necessitated cutting rough steps in the loose slope with our G-picks and groping our way upwards on all fours. The rim was quite wide and gently rounded so even in the strong wind we could stand there quite comfortably. However, due to the low cloud there was absolutely nothing to see, so there was nothing we could do except retrace our steps. The weather was getting worse, so rather than continue on to the other active cone of Benbow, we decided to head back to the shelter of our camp.

The following morning showed little improvement in the weather, so we packed up and headed back to the coast and our ship. By the time we reached the coast the weather had cleared. We continued our journey back to Santo where we arrived about 11 hours later.

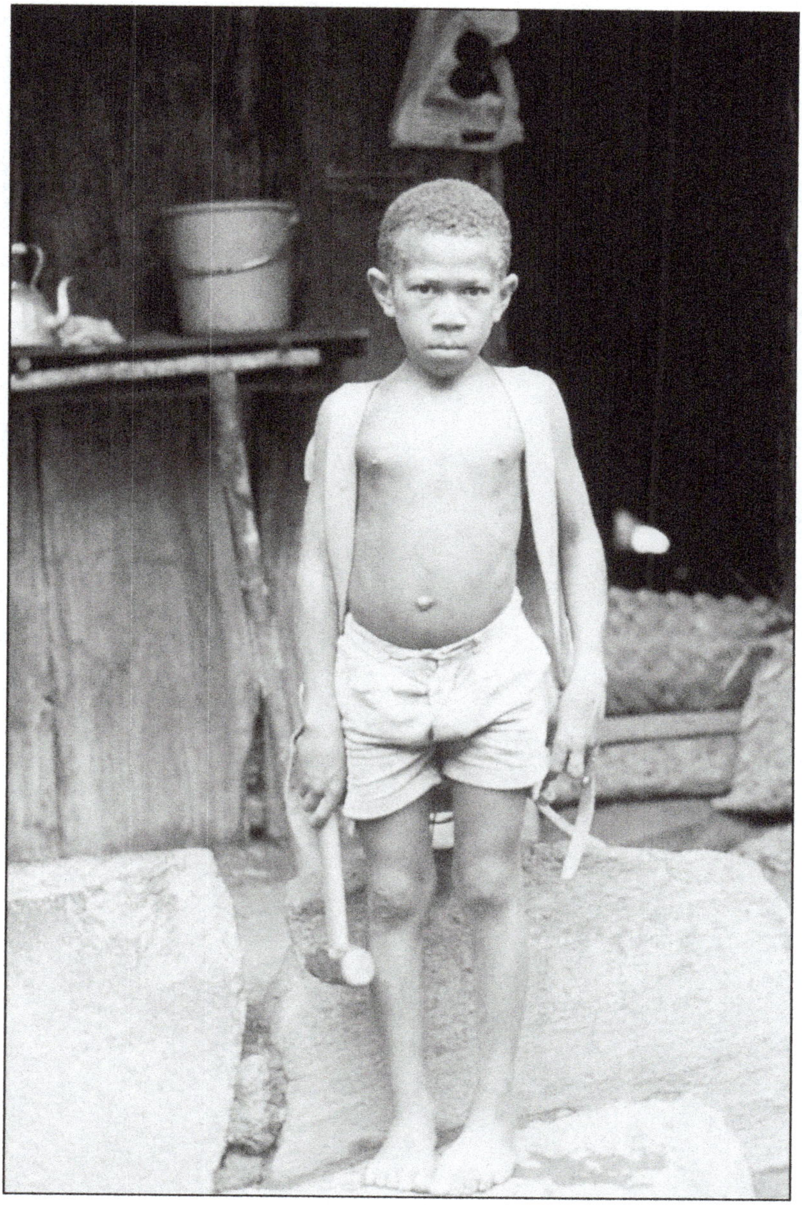

Above: Our assistant guide on the Merelava field trip.

NEW ZEALAND
Mt Ruapehu

Adventure to Mount Ruapehu

July 1975
New Zealand

I travelled to the Tongariro-Ruapehu National Park with Bruno, a friend of a friend, with the aim of climbing Mt Ruapehu. We headed off, armed with ice-axes, plenty of warm clothes and some food.

When we reached the Blythe hut track crossroads, we looked up at the mountain and perceived a rather interesting looking route up the mountain commencing with a large ridge on the other side of a wide valley. The usual route is to go up to the top of the Bruce, take the ski tows to a point not very far from the top, then follow an easy route to the crater rim. We thought *our* route promised more of a challenge so we headed across the valley and then slowly struggled up the ridge on the other side. The snow was very soft and pretty deep on this ridge so we spent quite a bit of time gaining the crest. Once on top of the ridge we followed its broad surface onto the mountain-proper. Until we stopped for lunch we were mostly plodding up gentle snow slopes. The scenery was magnificent with excellent views stretching away to Mt Egmont in the south-west and to Mt Ngauruhoe in the north. Lunch consisted of scroggin (peanuts, raisins and chocolate) and ice; our staple diet in the mountains.

After lunch we decided to follow a direct route up to the crater rim via a sharp ridge; we anticipated ascending a snow slope to the base of the ridge then climbing over the ridge via an easy-looking section. From there we would traverse onto another snow slope on the other side which we hoped would give us access to the summit via a longer, somewhat steeper snow slope beyond this. We made good progress to the ridge and reached the section we had observed as promising a way over the top. This section however proved to be an ice-slope covered with a thin veneer of snow followed by a nasty looking climb up a snow-covered rock projection. Bruno led the way up the ice, cutting steps with his axe. We rested amid a jumble of boulders to the side of this slope and contemplated the rock climb ahead of us. I led the way up this, spending a considerable amount of time clearing snow off the rock looking for handholds. Luckily there were just enough hand and footholds to get up and the rock proved very solid so no mishaps occurred. However, it was certainly not the way down. Bruno followed and we both felt quite relieved when we saw a soft snow slope on the other side of the ridge. This however led into a deep valley which we wished to avoid so we decided to stick to the ridge which continued right up to a point fairly near the crest of the rim. The rest of the ridge proved relatively easy, though slow as there were deep drifts piled up in places along its crest.

Near the highest part of the ridge a break in its continuation necessitated our following a snow traverse up a very steep looking snow slope. As the slope became steeper, the snow cover became thinner and eventually we were cutting steps in solid ice again. By the time we had struggled to the top of this slope, we were pretty tired so we had a scroggin stop before continuing our slow ascent.

That ice slope was in fact the last hurdle and the final route to the crater rim was a very long and arduous snow slope but not technically difficult. When we finally crested the rim we found an impressive drop on the other side; what we were standing on was in actual fact the caldera rim and not the crater rim. If anything, we were higher than the crater rim on this side but there was quite a distance separating us from the real crater. The rim we were standing on was quite narrow and it appeared that we might have some trouble getting off. The time was 4:30pm.

We started moving down and around the mountain, heading for the top of the Bruce section; however some icy patches inclined at a very steep angle caused us to descend too far and we found our passage around the mountain halted by steep bluffs. Sunset found us moving downhill and navigating steep bluffs which kept hindering our progress. We were now aiming to get onto the steep ridge containing the Whakapapaiti track and so make our way back to the hut. In the dim light we found walking on snow exceedingly difficult as a soft section or a drop would cause us to lose our balance and often precipitate a fall. By this time I was getting very tired and had to stop for frequent rests. Bruno however was like a work-horse and could have gone all night without a rest. For someone not quite 22, he showed amazing endurance. Possibly he was a lot fitter than I, but whatever the reason, I was glad of his company at that time. If we had decided to bivouac we would have been quite comfortable as the weather was excellent – no wind and very little cloud. However we kept on going, constantly getting stopped by bluffs, and finding a way down and around them. On one section, Bruno found the way down one of these by accidentally bum-sliding over a six-foot drop onto snow at the bottom. After his warning I was able to descend this bit more leisurely by using my ice-axe as a break. There was enough light from reflection off the snow to guide us and eventually we found ourselves on the ridge we wanted. This slope however proved the diciest of the lot as buried grass and boulders had produced a very uneven slope and we were constantly falling over. It was only a question of time though and by 8:30pm, after 12 hours on the mountain, we walked into the hut. The following morning, Bruno arose at 6:30 to meet some other trampers to do some more tramping and climbing. I enjoyed a leisurely breakfast around 9am before packing my things and heading on the return journey.

Postscript: Bruno was killed while rock-climbing alone in the Lakes District of England a few years later.

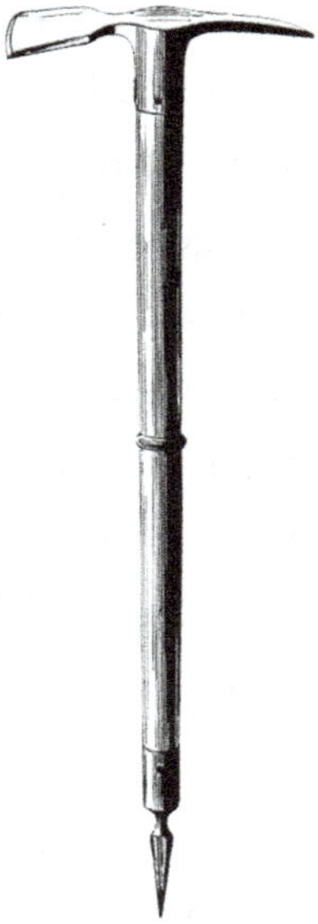

Vanuatu: Part II

May 1976

Second Banks Islands Field Trip

This second field trip to the Banks Islands was designed to complete the sampling work for my PhD thesis. This time we planned to visit Ureparapara (up to seven days), Merig (one day), Moto (three days), Motalava (seven days) and Gaua (three-plus days). I took a female field assistant (Anne) with me – another friend of a friend who was very keen to come. The trip was merged into a planned geological mapping program for the north Banks Islands by the resident geologists (Tim was the Chief Geologist at the time). I was essentially piggy-backing on their program.

Merig island is only 900m in length but rises to 125m in height. Its area is only 0.41km^2. As we were exploring the island for the maximum range of volcanic rock types, one of our passengers on the Euphosyne (our transport vessel) found a large hornblendite xenolith which was a very exciting discovery. This I later described as a cumulate (crystals settling to the bottom of a magma chamber and then picked up by an ascending magma). On Moto and Motalava, the Euphrosyne dropped us on shore with camping equipment and picked us up at the appointed time, thus enabling the ship to go off on some other errands. On Moto we had only one camp, the island being small enough to traverse without necess-itating a second base. Motalava was somewhat larger and so we moved camp every night, working our way around the perimeter of the island. We would go and do our traverses while the porters shifted camp under the supervision of a field assistant. When we reached Ureparapara I decided to try and avoid seven days on the island as we had originally planned, as the island was now thickly covered in ferns and nangalat (very nasty stinging nettles) following a recent cyclone. Fortunately, after three days a rundown trader steamed into the harbour. We grabbed our gear and persuaded the captain to carry us as far as Gaua, his next port of call. Our plan was to complete all the sampling work on Gaua before the Euphrosyne returned so that we could depart with that boat and hitch a ride all the way to Vila, saving us a lot of time, money and hassle. The three days spent on Ureparapara was long enough to collect some samples and get an idea of the geology of the island.

When we arrived at Losolava the chief gave us the radio shack to use as our base and even brought over a table and a couple of chairs. The hut was remarkably clean with a coral detritus floor in place of the usual dirt. A number of the islanders gave us island food which I sparingly consumed, neither greatly liking nor having sufficient room to do justice to everything we received. By this time

Anne was becoming paranoid with the bush and refused any of the local food.

We spent about four days at Losolava village on the north-east coast of Gaua Island (also known as Santa Maria island) and then hired an outrigger canoe with an outboard motor to transfer our base to Ontar village on the south-west coast. It was also a convenient way of observing the geology of the island and a lot easier on the feet. The chief at Ontar proved just as hospitable as our former host, though our new quarters were infested with rats. Compensating for this was our discovery of a brand new 'shack-out-the-back', complete with a wooden seat. There was no flush mechanism of course but it was the next best thing.

We spent three days in the Ontar area and then returned to Losolava on the Saturday by the same canoe. Having seen the Euphrosyne steaming past south Gaua early on Saturday morning, I calculated that she would arrive at Losolava some time on the Monday. So first thing on Monday morning, Anne and I and five porters set off for the crater with the intention of setting up an advance base camp at the foot of the active volcanic cone. The path to the top was clearly defined but very muddy so after three hours trudging through the mud in pouring rain we arrived at the lake's edge covered in mud and very cold. The natives then set about building a bamboo raft to carry us across the lake. Their technique was very simple; they cut down the bamboo then tied about 20 lengths together in a simple bundle. They then carried several of these bundles down to the water's edge where they tied four of them side by side. They then tied another four on top of these to complete the raft. No fancy lattice-work as in conventional rafts. The whole process took about three hours, so by the time we were ready to depart for the other side of the lake it was late in the afternoon. Paddles were cut from saplings. The raft moved through the water incredibly slowly, taking an hour and a half to travel a distance of just over a mile. Fortunately we were assisted by a fairly strong tail wind; as it was, we didn't reach the other side until after dark and we were faced with the unpleasant prospect of setting up a makeshift camp in darkness and pouring rain. The three porters we had kept with us helped us erect the flysheet, then found shelter under a large banyan tree nearby. Normally they share the flysheet with us but that particular location was so bad that there was barely enough room for two people to sleep.

We set off again early the next morning, paddling close to the shore to avoid the stronger swell offshore. It was still raining so our view of the volcano was restricted to its fern-swathed base, a result of recent ash deposition and subsequent plant regeneration. As we were shortly to find out, that regeneration had created an impenetrable barrier, in places 8' (2.5m) high, composed entirely of ferns. We found an excellent campsite near a small beach, so after beaching the raft the camp was constructed and a fireplace made under an overhanging tree. When everything had been done, I set off with two of the porters to return to Losolava by overland route, the intention being to rendezvous with the Euphrosyne, finalise plans with Tim and sleep on board before returning to the lake. The overland route involved going around the edge of the lake, over the lip of the caldera and then down the other side. Although the porters knew the track existed, it took over an hour to find it, backtracking numerous times through the swamps and pouring rain. The natives mark their tracks by cutting or notching vegetation on both sides of the passage, so we were continually hunting for old lesions, often hidden by sprigs of new growth. When we did eventually find the 'main road', we moved extremely fast to cover as much distance as we could before darkness enveloped the bush. The bush filters a lot of the available light anyway, so after sunset it is nearly pitch black under the forest canopy. The younger porter and I gradually drew away from the other porter (named Fred) and by dusk we had left him far behind. By 6:30pm it was too dark to proceed without a torch so we waited for Fred to catch up to us so that the three of us might proceed with the aid of my torch. After waiting vainly for about 15 minutes, the two of us set off again and finally arrived back at the village around 7pm, five hours after our departure from our lake camp. It was still raining heavily and to my great disappointment I noted that the Euphrosyne still hadn't arrived. This put me in a rather difficult position because I had been banking on spending the night on board. With no facilities for cooking and only a wet stretcher to sleep on I was forced to ask the chief if I could cook my rice in his house. He and his wife were very good and not only fed me but gave me a mattress and a pillow as well. The mattress was sponge rubber and the pillow kapok or something related. While I was dining with the chief, Fred returned and dropped off my very wet stretcher. On noting his arrival I went over to his house to see how he had got on navigating in the dark. He appeared to have had little trouble and was in good spirits so I was quite relieved.

The next morning heralded more torrential rain and little prospect of the arrival of the Euphrosyne. If it was miserable in the village, it must have been extremely unpleasant for Anne and David (the third porter) on top of the island. They reported later that it had rained continuously since my departure. The boat finally did arrive on the following morning but the moment she dropped anchor, a severe rain squall let loose and it was nearly two hours before I could get back on board. Tim explained that the ship's radio had broken down shortly after passing Gaua the previous Saturday and so she had been forced to return to Santo for repairs. While I had been cooling my heels on Gaua, Tim had been stewing in a malaria-ridden village on Ureparapara, as part of his geological mapping survey of the Banks Islands. Anyway, we quickly made plans for the next 36 hours – the length of time before the Euphrosyne would have to return to Santo and Vila. I would set off immediately for the lake with Tony (one of Tim's field assistants) and numerous porters to carry my geothermal gear, and return to Losolava by mid-afternoon the following day.

Tony and I loaded the boat and then rowed to the shore where we transferred the gear to the radio shack for distribution into packs. Shortly afterwards we set off with five porters for the long haul to the camp. The path had deteriorated considerably since our last traverse and we were continually wading through swollen streams and knee-deep mud. On reaching the end of the path we quickly inflated the rubber boat, launched it and loaded up the gear. We kept two of the porters with us and sped across to the camp, arriving there about 4.30 pm. After a brief meal, Tony, David and I set off for Sladen Boiling Springs, while Anne and two of the porters went out in the rubber dinghy to sound (measure the depth of) the lake. The route to the hot springs took over an hour and was a decidedly unpleasant traverse. Much of the ground was steaming and highly altered to red and yellow ochre. No actual fumaroles were seen or any violently boiling pools, though it is quite possible that the latter do occur. I had ten minutes before it became really dark, so I took some quick temperature and pH measurements, collected a water sample from one of the boiling pools and estimated the volume of water flowing out of the thermal area. Knowing the temperature of the water and the volume of outflow, one can calculate the amount of heat given off by the thermal springs.

Our departure from those hot springs can only be described as a groping retreat. We had only one very weak torch between us and what little light remained in the sky was filtered out by the thick bush. Trying to boulder-hop the hot acid waters of the outflow stream in nearly pitch darkness was a despairing business. It was with great relief that we finally climbed out of the gorge and returned to the security of mud and undergrowth. The torch managed to keep going until we reached the camp around 7pm. There we found the one remaining porter sitting on the tent smoking a cigarette. It was Fred, and I had asked him to keep the fire going, so I was rather annoyed to see the fire out and everything dark, wet and cold. Tony got the lamp going and David managed to start the fire again. A few minutes later, Anne and the other two porters arrived back. They had completed about three soundings in different parts of the lake, attaining a maximum depth of about 50m. This contrasted strongly with the results of a much earlier visit (about 1929) when a constant depth of 99m was obtained. Considering our equipment and the conditions under which we were operating I was not able to attach too much reliability to Anne's data. Whatever the correct depth, the volume of water in the lake is quite immense, for a crater lake anyway. If that amount of water drained into the volcano, the steam generated would be sufficient to pulverise the whole island. A large earthquake or eruption could easily bring about that scenario.

We spent a very wet night huddled in wet sleeping bags beneath a dripping flysheet; my pillow that night was a plastic sample bottle. Most of the porters spent a fairly cold, uncomfortable night as well, with the result that two of them had mild fevers in the morning. I dispensed some nivaquine (anti-malarial) tablets and they seemed to manage alright. After breakfast I left Tony and all the porters except David to pack up the camp and set off with Anne and David to look at the hot springs on the north side of the lake. We paddled there in about 20 minutes and then spent half an hour on the edge of the lake examining the small hot seeps that trickled into the lake. The pH was identical with the lake water so it would appear that the seeps represent lake water or meteoric water that has been heated by a shallow body of hot magma, rather than containing a direct contribution from the magma itself, as is probably the case with the Sladen springs.

We returned to the camp, picked up two of the porters and paddled across to the side of the lake abutting the active volcanic cone. The two porters dropped us here to begin our ascent of the volcano and returned to the camp to transfer all the gear to the head of the path on the other side of the lake. When they had finished they would then paddle back across the lake to pick us up on the other side of the volcano. From our landing point we started slowly into the thick undergrowth which became progressively thicker away from the shore. One of the field assistants, Edward, was at the front, hacking a path for us through the wall of fern. He would hack away at the fern for a while, then throw his body onto it to flatten it – a painstakingly slow process. After inching our way through this tangle for over an hour and covering less than two hundred yards

(180m), I realised there was no way in the world we would reach the top and descend to our rendezvous point in time to make the Euphrosyne's deadline. So we returned to our landing point hoping that we might see some sign of the rubber boat on its journey across the lake. The lake was empty and we were running out of time so we headed back to camp overland. The progression inland from the lake involves pandanus swamp at the lake's edge which gradually merges into a mat of fern on higher ground. As the fern would have taken too long to traverse, we were forced to find a route through the swamps. Wherever possible we clung to the branches, for an unsupported foot tended to sink a long way into the mud. At one point I tried to walk around the edge of a thick tangle of branches in waist deep water but rapidly changed my mind when a submerged branch I was standing on moved away from under me. The eels in this lake grow to a girth of six inches (15cm)!

After floundering around in this tropical paradise for about an hour, we managed to arrive back at our camp, which was now thoroughly deserted except for the framework of our tent and the bamboo raft. The boys quickly made some paddles and then we were off again on another inexorably slow journey across to the other arm of the lake. All four of us paddled that craft and it still took us about three hours to reach the rendezvous point. As soon as we sighted the other craft, we advised them of our presence in such a manner as to leave no doubt of our anxiety. The bamboo raft was abandoned and the six of us paddled the rubber raft across to the original landing as fast as we could go. Even with five paddles biting the water it took us over an hour to reach the other side. The raft was carried up to the clearing and deflated and packed. Some of our other gear was also up there so we apportioned it out amongst the porters and myself and began the long trek back to Losolava. After four hours paddling, the struggle through the swamp and the debilitating effects of cold, damp and very little sleep the night before, everyone looked a little jaded and weary. By the time Anne and I reached the coast we were pretty wacked.

With great relief we noticed that the Euphrosyne was still anchored, although we found out a little later that the Captain wouldn't have given us much more time. We were over two hours late as it was. While we were waiting for the Euphrosyne's dinghy I paid off the porters and bid my farewells. I felt a bit sorry for my poor porters who obviously hadn't worked so hard in a long time. Of the 60 or so people in the village, fewer than a dozen had actually been up to the lake and fewer than six had been across to the other side of the lake.

We sailed immediately and reached Santo in the early hours of the morning. We stopped there for about two hours then sailed again for Vila where we arrived about 17 hours later after a harrowing night journey. Both Anne and I spent all night leaning over the rail which effectively shattered our already tenuous constitutions. As a result we spent two days in a hotel recuperating.

Above: The dense Vanuatuan jungle. Below: Campsite with running water.

Above: Lopevi volcano. Below: Marum cone, Ambrym volcano.

Above: A local hut, Merelava.   Below: Bamboo raft, Lake Letas, Gaua.

Veladero
Cerro Bonete Chico
ARGENTINA

Adventure to Veladero Volcano

December 2005 - Age 57
Argentina

The motivation behind my solo trip to the Argentine altiplano was partly a desire to prove to myself that I could survive on my own in a harsh environment and partly my passion for climbing mountains, especially volcanoes in high altitude environments. The difficulty of finding a partner to join me was the final catalyst in making the decision to go it alone. Friends either had family commitments or lacked the necessary financial resources to cover the cost of the expedition. It had been a long time since I had done anything adventurous. At 57, and after rupturing an Achilles tendon and an anterior cruciate ligament in separate incidents on tennis courts a few years earlier, I felt that I was running out of time to do something exciting. There was, however, the issue of risk. I would be on my own and would have no means of contacting anyone if I needed assistance. The destination was a very isolated spot, over 100kms from the nearest village. Tourists and mountaineers occasionally visited this area but one could spend weeks there and not see a soul. As it transpired, I was completely alone for the eight days I spent above 4,600m. I would not take any foolhardy risks and it was not a case of summit at all costs. I also had a GPS, compass, map and an expedition tent.

My flights took me from Melbourne to La Rioja via LA and Beunos Aires (with a brief stopover in Catamarca). Upon arriving I wandered around the town searching for a trolley to cart my equipment up on the altiplano. Eventually I found a very strong, welded aluminium trolley. That afternoon I caught a bus to Villa Union near where I was staying in a recently constructed hotel. It was quite an imposing structure, solid brick and quite large with an unfinished swimming pool at the back. There were very few guests so there was no problem getting a room. Shortly after getting there I had a visit from Andreas who ran a business taking tourists to Laguna Brava and other popular tourist spots. He was my transport to and from where my journey would begin. He and his French wife, Clare, questioned me about my intentions to see if they could accommodate my requirements and to make sure I wasn't a nutter bent on self-destruction. Having a five-season tent and a GPS helped to allay their fears of a crazy gringo getting lost and dying out in the desert.

The next day was spent buying supplies for the trip. Unfortunately the range of tinned and packet food was quite limited, which was a major failure of this trip as canned goods were heavy and nearly everything was unappetising. At high altitudes one's appetite diminishes, so food needs to be tasty. I ended up with sardines, tuna, bread, oats, rice and milo. Having poor quality food at high altitude has a major bearing on energy levels and potential weight loss. I also had about 17kgs of water and two litres of fuel for the stove. My water supply was contained in two one-gallon desert canteens and numerous plastic soft drink bottles plus an aluminium flask. One of the desert canteens and the aluminium flask eventually leaked whereas none of the soft drink bottles leaked at all.

We departed Villa Union for the acclimatisation shelter at Refugio Penon (3,600m) on Wednesday the 7th, arriving there after about three hours of driving. The trip took us past the small towns of Villa Castelli and Vinchina on a good sealed road. Vinchina, at an elevation of about 2,000m, has two hotels and an internet café, supermarkets and restaurants. After that, the road was unsealed and climbed a narrow gorge with numerous stream crossings. In this section we passed some wonderful exposures of steeply dipping shales and mudstones. At one point Andreas stopped the car and showed me a tiny dinosaur footprint in mudstone on the side of the road. After about 20kms the asphalt returned and we entered the small village of Jague where we had to register and pay park fees (for entering the Laguna Brava National Park). The main street had been excavated about 2-3m below ground level so that when it rained the water would be channelled away down this street, thereby avoiding flooding. After Jague there were stretches of good asphalt for approximately 35kms to Punta Del Agua, a former gold miners' settlement consisting of 10-12 newly built barracks. Beyond this settlement there was only an unsealed road all the way to Refugio Penon at 3,600m. On the way we noticed small groups of vicuna and guanacos grazing on the stunted bushes in the valleys and on the hillsides. Both animals are relatives of the llama but the guanacos have dark faces whereas the vicuna have light-coloured heads. Puma are considered to be common in this area but they are rarely seen.

The refugio (28° 28.545 S; 68° 50.283 W) was located in a small river valley, close to the road and about 50m from the stream. It was a round building made of stones that had been constructed in the 19th century to provide shelter for herders driving their animals across the high passes into Chile. There was also another, slightly more modern but quite crudely built hut that I called the kitchen hut. At Penon there was a pipe coming out of the hillside about 30m from the buildings that provided a constant supply of fresh water. Andreas and Clare then returned to Villa Union, promising to pick me up in three days' time for the final journey to base camp at 4,600m.

The temperature at Penon at 4:30pm was 20°C. The first evening there I couldn't change the feeder nut on my stove to accommodate kerosene fuel so I resorted to making an open fire in the open fireplace of the kitchen hut. There was enough dry grass and small branches around from the stunted bushes that carpeted the valley floor to get the fire going. However, my fire got a little too large and was soon licking the branches of the roof. Fortunately the roof didn't catch fire, just blackened a little.

The wind never seemed to take a break up there so I erected my tent at the back of the kitchen hut where there seemed to be at least some shelter. This didn't seem to make much difference and for most of the night I lay inside my tent listening to the walls flapping like crazy. Along with a slight headache from the altitude the result was a fairly sleepless night. The minimum temperature during the night was 11°C, a level that was a bit warm for my monster down sleeping bag. I had had the bag for about 33 years, having purchased it in the US in 1972. It was a major pain to carry but I knew it would keep me warm, whatever the temperature. The next morning a large group of local tourists came through heading to Fiambala and Catamarca, so I took the opportunity to ask for some help with my stove. One of them had a tool kit and managed to undo the nut that needed to be replaced. After that I was able to get my stove working without any problems. I resolved to go for a walk on the following day to relieve the boredom and get my fitness back up to scratch. So the next morning I set off up the hill on the other side of the creek. The surface was quite loose, having accumulated quite a lot of scree material from aeons of frost shattering on the upper slopes. Nearer the top the gradient increased, making it quite hard going. Numerous animal trails assisted the task and eventually I was able to get to the top, where there was a small cairn. The view from the top was quite impressive, especially to the north, with Veladero quite visible in the distance. The altitude at the cairn was 4,212m, approximately 600m above the refugio but still 400m below the altiplano region we would be driving to tomorrow. That night the wind miraculously eased, allowing me to get a decent night's sleep.

The next morning Andreas arrived and drove me the rest of the way to the altiplano. The route took us past Laguna Brava, a large shallow salt lake and favourite tourist destination, and two more refugios. At the last refugio (Veladero) we stopped for half an hour for a break and to stretch our legs. The distance between the Refugio Laguna Brava at the head of the lake and Refugio Veladero is about 11kms. From there we followed 4WD tracks on a sandy plain to base camp, a spot nestled between Veladero and Cerro Bonete Chico. Veladero was the more imposing mountain, despite being about 300m lower, as there was a considerable amount of snow near the summit and the mountain had a more typical volcano shape. Bonete is a huge mountain with an extensive summit plateau topped by a number of rock slab pyramids. Base camp was located at 28° 05.404 S; 68° 51.48 W. Surprisingly, there was no wind when Andreas dropped me there. Just after he dropped me another group came through on their way to the Corona del Inca crater at around 5,300m. To go that far, tours always travel in convoy with at least two 4WD vehicles. I chatted with one of the guys for a short while and we compared GPS altitudes – both came up with exactly the same height, 4,668m.

After everyone had left I erected the tent and had a bit of a look around. Bonete looked to be a long way away and there was little snow and ice visible on the lower slopes. By contrast, there were clearly ice penitentes available within a couple of hours walking in the direction of Veladero. It was also a lower peak so might be easier to get to the top. So I decided to head for Veladero the following morning rather than take the riskier and definitely longer route to Bonete. It was quite eerie being all alone up there on the high plateau of the Andes. There were no humans, no animals, no birds, just the desert and the wind and a few clouds. I had an early dinner then retired to my tent and a warm sleeping bag. Sundown was just after 8pm. During the night the temperature dropped to 0.9°C. I kept all my water bottles in the tent to stop the water from freezing, and kept a small bottle in my sleeping bag in case the temperature dropped well below freezing.

The next morning I took a load over towards Veladero, looking for a good spot to make the next camp site. A short way beyond the crescent of ice penitentes I found two large boulders of ignimbrite that had obviously been transported here from the Corona del Inca crater where aerial photos show mounds of ignimbrite that are widely scattered. The rock was slightly vesicular and contained small crystals and some rock fragments in a light-coloured matrix. The space between the boulders provided some shelter from the wind so it was a good spot to set up a stove. Unfortunately it wasn't a very good spot to pitch a tent so that was erected a little way away. There was a large block of melting snow near one of the boulders which provided me with a useful supply of fresh water. I put my cup under one of the drips and periodically emptied the cup into my desert canteen. I also used my g-pick to remove lumps of snow for melting in the billy.

After going back for a second load I was completely exhausted and decided to wait until the following day to collect the last load. Although it only took around an hour to make the journey one way, the weight of the pack and the altitude took its toll. The elevation at the new camp (camp 2) was 4,815m and the location was 28° 05.444S and 68° 53.87W. That night my desert canteen leaked and destroyed one of my toilet rolls which happened to be lying beside it. Fortunately I had another bog roll so it wasn't a disaster! From then on I kept that roll in a secure place, well away from water bottles.

13th and 14th December 2005

When I went back for the last load, I couldn't find the location so I had to resort to the GPS and the appropriate waypoint to locate it. The GPS pointed the way straight to it, much to my relief, and I was soon marching back to the 'kitchen' camp. I spent the rest of the day in camp collecting water and resting. The next day I did a reconnaissance trip towards Veladero to find the best route and to try and find a suitable location for the next camp site. The first part of the recce involved a steep climb up a long hill then a relatively steep descent into a small valley or gully which then led onto the lower slopes of the mountain. Parts of the descent and subsequent ascent were on very difficult terrain – large angular rocks all jumbled on top of each other, and highly unstable. This terrain was typical of the steeper slopes where frost shattering had led to considerable accumulation of angular rock fragments of varying sizes. Trying to cross this terrain with a heavy pack would be really asking for trouble so I had to navigate a more workable route through this area. After working out a viable route I returned to kitchen camp and sorted out gear for the next day's push onto the slopes of Veladero. As I only had three days left before the scheduled pick-up, my plan was to take enough cold food for just two nights, leaving the stove and kerosene behind to save weight. I also took a full desert canteen with me and knew that I could supplement this with ice from the penitentes.

15th December 2005

Hauling the pack up the hill the next morning took an enormous amount of energy and by the time I reached the top I was exhausted. The next stage was fairly easy until I reached a wide band of penitentes blocking further progress up the mountain. I decided to try and force a way through rather than take the long route around the ends. This again took every last ounce of energy and I had to rest a while on the other side before continuing. I was now on a long slope with no flat areas so I plodded slowly upwards looking for a spot that could be manipulated into a camp site. At 5,093m I found a likely spot that required some levelling but was definitely workable. After levelling a platform with the ice-axe I managed to erect the tent and get my gear stowed safely inside. The wind had been relatively mild on the way over but now shifted around to the south and picked up intensity. This meant that my tent was now facing the wrong way (the high end into the wind) but I wasn't about to dismantle it and re-erect it, so just hoped the tent would cope. Fortunately it did.

16th December 2005

'Summit' day dawned bright and clear as I faced the prospect of climbing 1,300m, an impossible task at this altitude. After four hours of climbing I reached 5,735m (642m higher than I had been). At this stage the terrain was deteriorating into a chaotic assemblage of shingle. I decided to call it quits at this point as I was very tired from effort and insufficient food and thought there was a real risk of an accident if I kept going. So I took some photos and then headed back down the mountain, reaching the tent an hour and a half later. I spent the night there and then returned to kitchen camp the following day. The winds were very strong that day so breaking camp was quite an ordeal. Back at kitchen camp I set up the tent in a new spot to try and get more protection from the wind. Then I melted some snow and dined on rice and sardines. After that I took a pack over to the rendezvous point so that I would only have one trip to make the following morning.

18th December 2005

The next morning I packed up and made the last trek to the rendezvous point, arriving there around 9:30am. I had just dropped everything and sat down on the ground when I noticed a plume of dust coming from the 'road' to the south. This was Andreas, in company with Clare and another man, so my timing was excellent. Shortly after we set off, I was offered a ham and cheese sandwich, a rare treat after two weeks of incredibly boring food. It was the best sandwich I have ever had. We travelled back to Villa Union where they dropped me back at my Hotel. Stepping into that hotel I felt like Colonel Fawcett returning from one of his adventurous travels. In fact there was no-one to impress as it transpired that I was the only resident!

After a wonderful shower I tried to shave off my two-week beard. The razor was a little blunt and made virtually no impact on my whiskers so I had to walk a kilometre to the nearest shop to buy a new razor. Eventually I was able to remove the growth and feel like a new man. At dinner time I walked into the dining room and found just one place made up – mine! That night I had steak and chips, washed down with a large bottle of beer, and to Hell with the low cholesterol diet.

Adventure to Cerro Bonete Chico

November 2008 - Age 60
Argentina

This time I travelled with a friend, Graham (age 59). The plan was to try and climb Cerro Bonete Chico, a near neighbour of Veladero and, at 6,759m, the fourth highest mountain in South America. To overcome the food problems I had experienced previously, I took 8kgs of packaged food with me – the allowable limit. This food comprised energy bars, quick oats, powdered milk, Sustagen, sultanas, soup packets, mixed nuts and enough freeze-dried dinners to last the distance.

29th November 2008

Our bus departed La Rioja around 2pm and arrived at Villa Union at 6pm. After breakfast the next day we boiled a dozen eggs in the hostel kitchen so we could take half a dozen each to cover the first week or so of lunches. At 10am Andreas turned up in the 4WD with our kerosene and we took off for Refugio Penon, our acclimatisation camp. The trip took 3.5 hours so when we got there we got the burner going and had soup and bread and an apple. The rest of the afternoon was occupied in setting up the tents and filling our water bottles from the pipe in the hillside.

1st December 2008

During the evening and night the temperature inside the tent fell from 28°C before dusk to 10.6°C. The outside air temperature at 7:30am the next morning was 9°C. The wind dropped in the early morning which was a nice respite, if short-lived. For most of the night I had a slight headache from the altitude but this went by 8:30pm. After breakfast Graham and I went up the nearby hill, a two-hour slog on loose rock. The top was at 4,215m and gave a clear view to Veladero in the north.

That evening I was sitting on Graham's three-legged stool when it collapsed under me and I fell back and hit my back on the edge of the concrete slab that formed our table. Fortunately it didn't appear to have done any damage apart from some bruising. That afternoon we tried to have a rest in the tents but found them insufferably hot (up to 35°C). This would be one of the main complaints of the trip; daytime temperatures inside the tents were too hot for any sort of comfort and if we opened the flaps (front and back) to let some air through, the tent filled up with fine sand. For old bastards like us, it was a real problem. I went to bed about 5:30pm and lay on top of the sleeping bag until the sun went down and provided us with some relief from the heat. That night the temperature dropped to zero outside the tent but only fell to 3.8°C inside my tent.

The next day we tackled the big mountain on the other side (western side) of the valley. We departed at 9:30am and reached 4,200m at 12:20pm. At this point we were pretty stuffed and were still well short of the top so we had lunch then headed back to camp. The terrain at our lunch stop was very sandy so it had been quite hard going. That afternoon the temperature inside my tent reached 35.5°C.

3rd December 2008

Andreas picked us up around 11am and took us up to our base camp at 4,670m (location 28.05.65S and 68 51.3W; waypoint 6). It was blowing a gale and quite cold. The wind made it quite difficult to get Graham's tent up as it was not terribly stable in high winds. We spent the rest of the afternoon building a rock wall around the windward side of the tents to provide some protection for the stove and tents. We only used the stoves outside the tent as I recalled the nasty experience with an out of control kerosene burner in my tent in Antarctica. During the night it got down to -5.7°C in my tent.

The next day we did a carry to the next camp site, following a dry stream bed. The journey was arduous as we were walking on sand for much of the time and had to negotiate one high sand dune in the middle of the river valley. The camp site was reasonably sheltered with water from a small quebrada (rivulet) caused by melting of a nearby field of penitentes. The altitude was 4,845m. We called this advance base camp and the location was 28° 04.712S and 68° 50.267W (waypoint 7). After selecting the site and depositing the bulk of our food and fuel, we returned to base camp. The next day we packed up the tents, sleeping gear, stoves and other essential gear and headed for advance base camp. Once there we set up then rested in the sweltering tents. As the water in the stream was coming directly from runoff from melting penitentes we filled our water bottles without boiling it first, and had no ill effects.

6th December 2008

We did a carry to Camp 1 further up the river valley. Disconcertingly, we found tyre tracks from a point further east that led right up to this camp site at 5,111m, indicating that some climbers had managed to get here without an arduous four-day trek. This was quite demoralising as the longer one is out here the more one is worn down by the conditions. The constant wind and sweltering tents meant there was no escape, so by the end of day four we were already worn out mentally. The location of this camp was 28° 03.873S and 68° 49.148W. When we had set up the tents, Graham said we should try and find a faster way of moving up the mountain. I had been thinking about this too so we discussed it over dinner. In the end we decided to do a reconnaissance trek without gear to see where we could pitch a final camp. So the next day we set off with just our lunch and some water and headed up directly towards the mountain and slightly away from the river channel. It was pretty tough going but we eventually came to a relatively flat area where we thought we could make a reasonable camp at around 5,500m. Unfortunately there wasn't any water there and it was still 1,200m below the summit so we knew we would not make the top. Still, we thought we could make 6,000m which would be a new high for both of us. So we returned to our camp, this time along the river bed which was easier, although there was one quite steep section.

After dinner we contacted Andreas on the satellite phone and told them we planned to be back at base camp on the 11th. We arranged to be picked up that day around 2pm at the location where Andreas had dropped us off. This would be relatively easy for us to find as it was where we had built the stone wall.

9th December 2008

The next day we set off on the direct route, avoiding the river channel. This was actually a mistake as it proved to be totally exhausting. It was not so much the steepness as the nature of the terrain, which was carpeted with angular blocks and made walking quite difficult. The river channel would have been easier, even with the steep section. By the time we reached our proposed camp site we were completely buggered so had an early dinner and retired to bed.

The next morning we set off reasonably early (around 8am) for our last push. We headed straight up a very steep slope which was quite arduous. In fact the whole climb that morning was quite steep on a surface that was relatively loose. We eventually reached 6,000m and decided to call it quits as it was getting too steep to go much higher. We would have had to move laterally and then descend a little before heading up again. By this time we were well and truly stuffed so were quite prepared to retreat. We headed back down to the tent where we packed everything up and took off for Camp 1, arriving there late in the afternoon.

11th December 2008

We packed up and left camp reasonably early as we wanted to be sure we got back to base camp before Andreas arrived. We had some trouble finding the camp site but got there eventually with the aid of my GPS. We dumped our bags, had lunch and waited. And waited. Around 4pm we noticed two cars a long way off that seemed to be waiting for something or somebody, but they were so far away we couldn't believe it was Andreas. We felt it was too far to go hiking over there to find out and so we assumed it was another party and waited for the 'real' Andreas to turn up. Eventually we realised no-one was coming that day and that we would have to erect the tents again and resume camping. The last thing we wanted to do was put the tents up again and then try to light the stoves to make dinner. We were low on food and there was only enough water for two days. That evening we tried to ring Andreas but couldn't get through.

The next day no vehicle turned up which caused us to speculate that they might have had an accident and hadn't been able to arrange for someone else to collect us. That afternoon we walked across to a penitente field and collected some ice for melting to supplement our water supply. Waiting around for someone to turn up was painful in the extreme; inside the tents it was boiling hot, and outside we were getting sand-blasted. As the tents were so hot we were able to melt a lot of the ice just by storing it inside in waterproof receptacles.

Two days after our planned departure we spotted Andreas' vehicle searching some distance away and managed to chase it down. Andreas said that the two cars we saw two days ago had been his group.

Editor's note:

These were the last words in Mark's expedition diaries.

Mark Barsdell

Above: Transport car at drop-off point. Below: Scree terrain.

Above: The beginning of nowhere. Below: The stone refugios.

Epilogue

by Jeremy Majid

The trip to Argentina in 2008 would become Mark's final journey to the most remote parts of the earth. It brought to a close a series of travels that can genuinely be given the term explorations; for he and those he worked with were bringing back with them new information about our planet. For much of his time away he endured a level of isolation that is almost hard to fathom in today's ultra-connected society. And at the age of 60, well entrenched in the modern office culture, he still found the mental stamina and physical capability to continue to push new boundaries.

This is not to say that he has given up his love of volcanoes, nor the outdoors. Now in his 70s, he is well entrenched in a physically active retirement; one wonders how he ever had time for full-time work. He and his wife Susan moved to a small hobby farm on the south east coast of Australia where he indulges his other great passion: trees. After several years on the farm, Mark has worked tirelessly to cultivate a range of flora, much of which is edible, including sizeable crops of blueberries, and increasingly, avocadoes. He still has an eye to the future, and has planted a range of young trees including Japanese Maples and unusual pine varieties, which are lovingly tended. And while many decades have passed since Mark's involvement in academia, he still keeps meticulous records of the annual rainfall and holds a keen eye on climate change literature.

The ice-axe and crampons are no longer required, but that doesn't mean Mark and Susan have lost the travel bug, with recent trips taking them to Mt Vesuvius (Pompei), Hawaii and Mt Fuji; one can never see too many volcanoes. They also regularly visit their son Ben in the United States, and their daughter Wendy and her family in Adelaide, Australia.

Made in the USA
Monee, IL
04 November 2026

67387011R00118